SHAPE MEMORY ALLOYS - FUNDAMENTALS AND APPLICATIONS

Edited by **Farzad Ebrahim**

Shape Memory Alloys - Fundamentals and Applications
http://dx.doi.org/10.5772/66914
Edited by Farzad Ebrahim

Contributors

Takumi Kihara, Xiao Xu, Ryosuke Kainuma, Wataru Ito, Yoshiya Adachi, Takeshi Kanomata, Masashi Tokunaga, Takashi Suzuki, Shuji Yoshikawa, Shinya Okabe, Xuexi Zhang, Mingfang Qian, Elena O. Nasakina, Sevostyanov Michail, Alexander S. Baikin, Sergey V. Konushkin, Konstantin V. Sergienko, Alexander V. Leonov, Alexey G. Kolmakov, Alexey V. Seryogin, Jawaid Daudpoto, Atta Muhammad Nizamani, Muhammad Ali Nizamani

Published by InTech
Janeza Trdine 9, 51000 Rijeka, Croatia

Publishing Process Manager Romina Rovan
Technical Editor SPi Global
Cover InTech Design team

First published September, 2017
Printed in Croatia
Legal deposit, Croatia: National and University Library in Zagreb

Additional hard copies can be obtained from orders@intechopen.com

Shape Memory Alloys - Fundamentals and Applications, Edited by Farzad Ebrahim
p. cm.
Print ISBN 978-953-51-3455-8
Online ISBN 978-953-51-3456-5

For EU product safety concerns:
IN TECH d.o.o., Prolaz Marije Krucifikse Kozulić 3, 51000 Rijeka, Croatia,
info@intechopen.com or visit our website at intechopen.com.

Meet the editor

Dr. Farzad Ebrahimi received his MSc and PhD degrees in Mechanical Engineering, with a specialization in applied design, from the University of Tehran, Iran, in 2009. He joined the department of mechanical engineering of Imam Khomeini International University as an assistant professor in 2010.

He is involved in several international journals as editor and reviewer. He serves on the editorial board of the "Journal of advances in mechanical engineering." He is the author of books *Smart FG Plates* and *Progress in Analysis of FG Structures*, Nova Science Publishers, NY. He also served as the editor of the book *Nanocomposites - New Trends and Developments*, InTechOpen Publisher. His research interests focus on areas of smart materials and structures, nanostructures, vibration nanocomposites, and composite materials and structures, and he has published several researches in these fields.

Contents

Preface

Shape memory alloys (SMAs) are metals that "remember" their original shapes and exhibit two very unique properties: pseudo-elasticity and the shape memory effect. SMAs have drawn significant attention and interest in recent years in a broad range of commercial applications, due to their unique and superior properties; this commercial development has been supported by fundamental and applied research studies.

This book is a result of contributions of experts from international scientific community working in different aspects of shape memory alloys and reports on the state-of-the-art research and development findings on this topic through original and innovative research studies. Through its five chapters, the reader will have access to works related to ferromagnetic SMAs, while it introduces some specific applications like development of faster SMA actuators and application of nanostructural SMAs in medical devices.

The book contains up-to-date publications of leading experts, and the edition is intended to furnish valuable recent information to the professionals involved in shape memory alloys' analysis and applications. The text is addressed not only to researchers but also to professional engineers, students, and other experts in a variety of disciplines, both academic and industrial, seeking to gain a better understanding of what has been done in the field recently and what kind of open problems are in this area.

I hope that readers will find the book useful and inspiring by examining the recent developments in shape memory alloys.

Lastly, I would like to thank all the authors for their excellent contributions in different areas covered by this book and the InTechOpen team, especially the publishing process manager Ms. Romina Rovan, for their support and patience during the whole process of creating this book.

Dr. Farzad Ebrahimi
Department of Mechanical Engineering
Imam Khomeini International University,
Qazvin, Iran

Section 1

Shape Memory Alloys: Fundamentals

Chapter 1

Ferromagnetic Shape Memory Alloys: Foams and Microwires

Xuexi Zhang and Mingfang Qian

Additional information is available at the end of the chapter

http://dx.doi.org/10.5772/intechopen.69702

Abstract

Ferromagnetic shape memory alloys exhibit martensite transformation (MT) and magnetic transition and thus may be actuated by thermal and magnetic fields. The working frequency of these alloys may be higher than conventional shape memory alloys, such as Ni-Ti, because the magnetic field may operate at higher frequency. This chapter focuses on some fundamental topics of these multifunctional materials, including the composition-structure relationship, the synthesis of the foams and microwires, the martensite transformation and magnetic transition characters, the properties (magnetic-field-induced strain (MFIS), magnetocaloric effects (MCEs), shape memory effects, and superelastic effects), and applications. The improvement of the magnetic-field-induced strain due to the reduced constraint of twin boundary motion caused by grain boundaries in polycrystalline Ni-Mn-Ga foams and the size effects of the superelasticity and magnetocaloric properties in Ni-Mn-X (X = In, Sn, Sb) microwires are detailed and addressed.

Keywords: ferromagnetic shape memory alloys (FSMAs), Ni-Mn-Ga alloys, foams, microwires, martensite transformation, ferromagnetic-field-induced strain (MFIS), magnetocaloric effects (MCEs)

1. Introduction

Ferromagnetic shape memory Heusler alloys, such as Ni-Mn-Ga and Ni-Mn-X (X = In, Sn, Sb), are receiving increasing attentions due to their multifunctional properties, that is, magnetic-field-induced strain (MFIS), magnetocaloric effect (MCE), magnetoresistance, etc. Before 1996, some works concern the martensite transformation (MT) of Ni-Mn-Ga alloys [1–3]. The relationship between the composition on martensite and magnetic transformation temperatures has also been revealed [4]. In 1996, Ullakko et al. published the first paper on the MFIS of single-crystalline Ni-Mn-Ga alloys [5]. Since then, high MFIS of 6.4% has been found in Ni-Mn-Ga alloys with 5M martensite structure [6] and of 10% with 5M martensite structure [7]. By allying a high magnetic field, the single-crystalline Ni-Mn-Ga alloy with non-modulate (NM) martensite structure may show a giant MFIS as high as 12% [8]. In 2007, Müllner et al. published their first paper on Ni-Mn-Ga foam and found that 0.24% MFIS may be achieved in the foams with single-model pore architecture [9]. By introducing secondary pores, thus forming dual-pore architecture, the foams may exhibit a MFIS of 8.7% after suitable thermal-magnetic training, which approaches the theoretical limit of single-crystalline Ni-Mn-Ga alloys with 14M martensite structure [10]. In 2008, Scheerbaum et al. [11] produced Ni-Mn-Ga fibers with diameter ~60–100 μm, which showed a 1% MFIS upon applying an external magnetic field of 2 T. Since 2012, researchers in Harbin Institute of Technology make use of the so-called melt extraction method to fabricate microwires on a large quantity [12, 13]. The obtained microwire has a naked surface, which can be directly used in MEMS or NEMS, and acts as building blocks for composites and complex-shaped components. Systematic research works have also been carried out to reveal the superelasticity [14, 15], shape memory [12, 16], and MCE properties [17–19].

Ni-Mn-X (X = In, Sn, Sb) alloys, another important ferromagnetic shape memory alloy family, attracted attention of the scientific society since Kainuma et al. [20] reported the giant stress output in Ni-Mn-In-Co alloys. On contrary to Ni-Mn-Ga alloys, these Ni-Mn-X alloys exhibit paramagnetic/antiferromagnetic martensite and ferromagnetic austenite. As a result, a bias magnetic field may stabilize the austenite phase, which is responsible for the shift of martensite transformation temperatures to lower temperature. Such metamagnetic structural transition from the paramagnetic/antiferromagnetic martensite to ferromagnetic austenite under a bias magnetic field may produce giant MCE [21].

This chapter describes the synthesis, processing, and properties (especially the MIFS and MCE) of these alloys, with emphasis on Ni-Mn-Ga foams and microwires. The up-to-date results, mainly reported in the past decades, will be covered. The aim of this chapter is to provide researchers an overview of the background, current status, and future development of the ferromagnetic shape memory alloys.

2. Composition-structure relationship in ferromagnetic shape memory alloys

2.1. Relationship between composition (*e*/*a*) and transformation temperatures

Off-stoichiometric Ni-Mn-Ga ferromagnetic shape memory alloys (FSMAs) have been attracted much attention because of their multiplicity of functional properties like excellent magnetic-field-induced strains (MFIS) [6, 7, 10], magnetocaloric effect (MCE) [22–24], conventional/magnetic shape memory effects (SME) [12], etc. The martensite phase or the martensite transformation (MT) temperatures (martensite start and finish temperature M_s, M_f, and the austenite start and finish temperature A_s, A_f) of the alloys are closely related to their functional properties. For instance, the Ni-Mn-Ga alloys show MFIS only in the martensitic state owing to the reorientation of the twin boundaries in an applied magnetic field. As for the MCE, the effective refrigeration working temperature interval (WTI) of the cooling system is within the MT range. For Ni-Mn-Ga alloys, the MT temperatures of the alloy are particularly sensitive to their compositions [4].

Here, we focused our content on Ni-Mn-Ga alloys due to the space limit. For the stoichiometric Ni_2MnGa, the MT temperatures are lower than room temperature (RT), ~185 K; thus, no martensite exists at RT. By varying the composition of 5 at.%, a huge M_s change from 154 to 458 K can be obtained [4, 25–30]. The effect of composition on MT temperatures of the Ni-Mn-Ga alloys was studied by Chernenko et al. at 1995 [4]. A general view of transformation temperature shifting as a function of the Ni/Mn/Ga element content was summarized as (1) at a constant value of Mn content, Ga addition lowers M_s temperature, (2) Mn addition (instead of Ga) at constant Ni concentration increases M_s, and (3) substitution of Ni atoms by Mn at constant Ga content results in alloys with lower M_s.

Furthermore, based on the M_s and transformation enthalpy (ΔH), Ni-Mn-Ga alloys were classified into three groups, as demonstrated in **Table 1** [4, 31, 32], where T_c stands for the magnetic transformation of Curie temperature of the alloy.

Thereafter, more systematic and quantized studies have been performed [25–30]. The effects of composition on temperature and ΔH can be formulated by the linear regression listed as follows [30]:

$$M_s\ (\mathrm{K}) = 25.44\mathrm{Ni}\ (\mathrm{at.\%}) - 4.86\mathrm{Mn}\ (\mathrm{at.\%}) - 38.83\mathrm{Ga}\ (\mathrm{at.\%}) \quad (1)$$

$$\Delta H\ (\mathrm{J/g}) = 0.72\mathrm{Ni}\ (\mathrm{at.\%}) - 0.16\mathrm{Mn}\ (\mathrm{at.\%}) - 1.23\mathrm{Ga}\ (\mathrm{at.\%}) \quad (2)$$

The number of valence electrons per atom (e/a) for Ni, Mn, and Ga atoms is $10(3d^{9.9}4s^{0.1})$, $7(3d^5 4s^2)$, and $10(4s^2 4p^1)$, respectively [26]; thus, the valence electron concentration of the alloy can be defined as

$$e/a = \frac{10\,\mathrm{Ni}_{at\%} + 7\,\mathrm{Mn}_{at\%} + 3\,\mathrm{Ga}_{at\%}}{\mathrm{Ni}_{at\%} + \mathrm{Mn}_{at\%} + \mathrm{Ga}_{at\%}} \quad (3)$$

The relationship between M_s temperature and e/a is summarized and presented in **Figure 1**. Statistical analysis shows a relatively large standard deviation of 44.7 K and a maximum error of 140.2 K for the extreme case, which implies that the MT may also be very sensitive to the microstructure of the alloy, that is, internal stress and the degree of the atomic order [14].

Group	e/a	M_s (T)	ΔH (J/g)
Group I	<7.7	$M_s < RT < T_c$	~1.6
Group II	7.55–7.7	$M_s \approx RT < T_c$	~4.2
Group III	>7.7	$M_s > T_c$	~8.5

Table 1. Classification of Ni-Mn-Ga alloys according to M_s and enthalpy (ΔH) [4, 17].

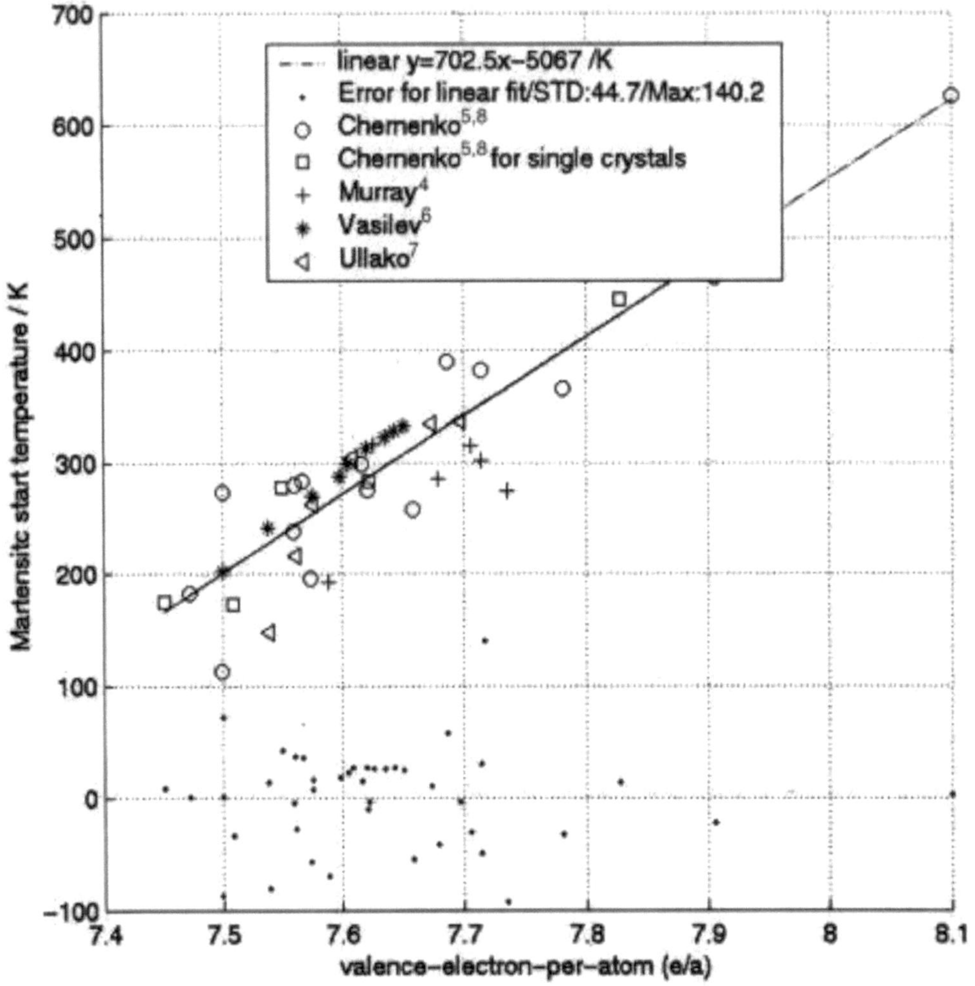

Figure 1. The relationship between martensite transformation temperatures and valence electron concentration of Ni-Mn-Ga alloy [29].

Based on **Figure 1**, the effects of *e*/*a* on M_s temperature can be formulated by the linear regression, as listed in Eq. (4), indicating that the MT temperature increases with increasing *e*/*a*:

$$M_s = 702.5(e/a) - 5067 \tag{4}$$

Besides, T_c varies less with composition variation (Mn, 20–35 at.%; Ga, 16–17 at.%) than M_s for Ni-Mn-Ga alloys [17, 29]. The effect of composition on the T_c of Ni-Mn-Ga alloy mainly relies on the Mn-Mn distance since the ferromagnetism mainly depends on the Mn-Mn atomic interaction [33]. The stoichiometric Ni_2MnGa alloy possesses the strongest ferromagnetic interaction between the neighboring Mn-Mn atoms with T_c ~ 376 K. Either with the concentration or expansion of the unit cell, the Mn-Mn nearest distance may be decreased; thus, the ferromagnetic interaction between neighboring Mn-Mn atoms can be weakened, which gives rise to a slight decrease of T_c [34].

Above all, the tunable transformation temperatures lead to multifunctional properties in Ni-Mn-Ga alloys, which will be demonstrated in details in Section 5.

2.2. Phase and structure of the martensite

Ni-Mn-Ga alloys exhibit a chemical ordering transition (a kind of second-order phase transition) from partially ordered high-temperature B2 phase to chemical ordered $L2_1$ phase during cooling. In the $L2_1$ phase, the Ni, Mn, and Ga atoms occupy the specific sites of the crystal lattice, as shown in **Figure 2**.

The $L2_1$ phase has a cubic face-centered lattice. On the other hand, Ni_2MnGa alloy displays martensite transformation (a kind of first-order phase transition). The martensite phase may show various stacking sequences, that is, a modulate structure, which thus creates martensite structure, such as five-layer modulate (5M) martensite phase (**Figure 3**), seven-layer modulate martensite phase (7M, **Figure 4**), and non-modulate (NM) martensite phase (**Figure 5**). The kind of modulation mainly depends on the composition of the alloy. These different martensite structures exhibit different twinning stresses, which has a large effect on the magnetic-field-induced strain and will be summarized in detail in Section 5.1.2.

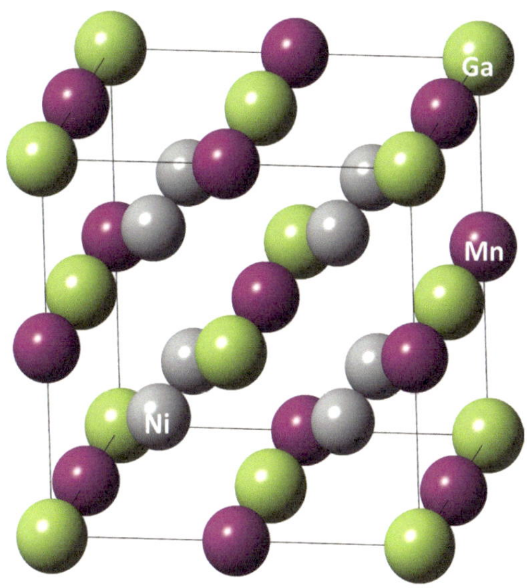

Figure 2. $L2_1$ crystal structure of the Ni_2MnGa alloys.

3. Synthesis of ferromagnetic Ni-Mn-Ga foams and microwires

3.1. Ni-Mn-Ga foams

Ferromagnetic Ni-Mn-Ga foams can produce large magnetic-field-induced strain of ~2.0–8.7% due to the reduction of constraints imposed by grain boundaries and the formation of bamboo

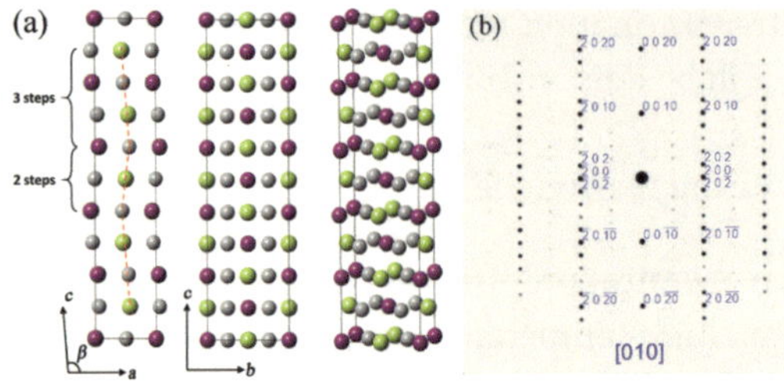

Figure 3. (a) 5M crystal structure of the Ni_2MnGa alloys and (b) simulated diffraction pattern of 5M martensite along crystal zone axis of [0 1 0].

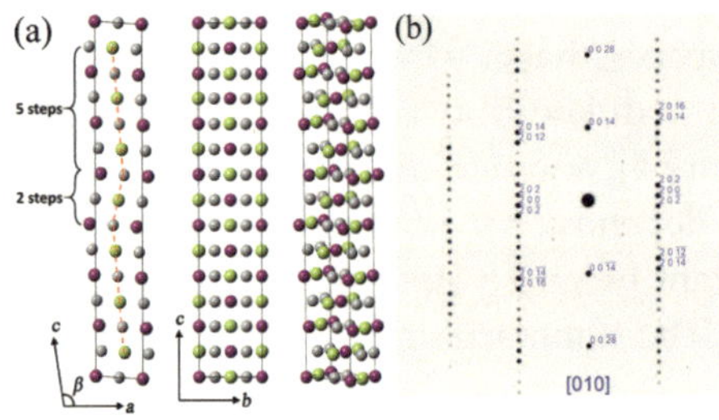

Figure 4. (a) 7M crystal structure of the Ni_2MnGa alloys and (b) simulated diffraction pattern of 7M martensite along crystal zone axis of [0 1 0].

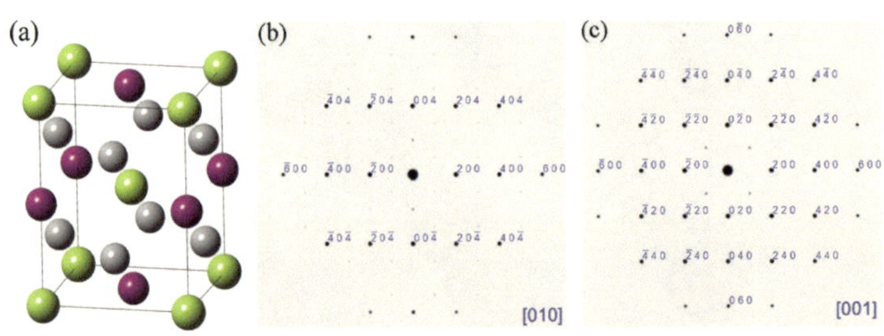

Figure 5. (a) NM crystal structure of the Ni_2MnGa alloys and simulated diffraction pattern of NM martensite along crystal zone axis of (b) [0 1 0] and (c) [0 0 1].

grains in the struts [10]. In this section, the synthesis of the Ni-Mn-Ga foams with single-pore and dual-pore distribution is demonstrated in detail [9, 35].

The casting replication method, that is, using liquid metal infiltration of a preform of ceramic space holder powder [36, 37], was used to prepare the Ni-Mn-Ga foams. Sodium aluminate ($NaAlO_2$) was used as space holder in this case due to its high melting temperature ~1650°C, excellent chemical stability with molten metals, and good solubility in acid. In order to create the single-pore and dual-pore distribution, different sizes of $NaAlO_2$ powders were prepared as follows: (1) purchased $NaAlO_2$ powders was cold pressed at 125 MPa and sintered at 1500°C for 3 h in the air; (2) the sintered body was broken up with a mortar and pestle into powders; and (3) the resulting powder was sieved into different size ranges: R1 for single-pore foam and R2 (coarse) and R3 (fine) for dual-pore foam. The specific size of the powder was tuned due to the requirement of the foam porosity.

For single-pore specimen, the R1 powder was directly poured into an alumina crucible and slightly trapped to the designed height. For dual-pore specimen, the coarse and fine powders were poured alternatively (layer by layer) in the crucible filled with acetone: the coarse powder was first poured in a small batch, followed by a small batch of fine powder to settle in the space between the coarse ones of the previous batch. After that, the crucible was heated to evaporate the acetone. Both crucibles were heated at 1500°C for 3 h in the air to create necks between powders for the infiltration of the Ni-Mn-Ga liquid metal and the formation of open pores.

For infiltration process, the ingot was placed on top of the sintered powder preform and then heated to melt the Ni-Mn-Ga alloy under vacuum in the furnace. After melting the alloy, high-purity Ar gas was introduced in the furnace at a pressure of 1.34 atm. to squeeze the molten alloy into the preform, and the temperature was then cooled to room temperature at 7°C/min.

The removal of the $NaAlO_2$ is critical for the preparation of the Ni-Mn-Ga foams. For the single-pore foam, a 10% HCl solution was used. Ten percent of HCl solution not only solve the $NaAlO_2$ but also solve the Ni-Mn-Ga alloy, thus, thinned the struts. In this case, for the dual-pore foam, the thin struts around the fine powders would be dissolved after long exposure to the 10% HCl solution. Thus, a two-step method was applied: (1) removal of the coarse powders without alloy dissolution and (2) removal of the remaining fine powders as well as some thinning of the alloy. The acid for the first step should dissolve $NaAlO_2$ but not the alloy. The acid for the second step should rapidly dissolve $NaAlO_2$ while slowly dissolving the alloy only to open small fenestrations between the fine $NaAlO_2$ powders. The mass loss of the bulk nonporous Ni-Mn-Ga alloy vs. time plots is shown in **Figure 6a**, which are linear for the acids of 10% HCl, 20% HCl, and 34% H_2SO_4. The slopes corresponding to dissolution rates are 16 and 5 mg/m^2 min in 10% HCl and 34% H_2SO_4, respectively, while triple the rate when doubled the concentration to 20% HCl.

As a result, 34% H_2SO_4 and 10% HCl were selected for the two steps, respectively. **Figure 6b** plots the mass loss vs. time for a foam sample with dual-size $NaAlO_2$ powder immersed in 34% H_2SO_4. After 2000 min, only 87% of the $NaAlO_2$ was removed from the sample. Thereafter, a similar foam sample was firstly immersed in a 34% H_2SO_4 solution for 645 min (corresponding

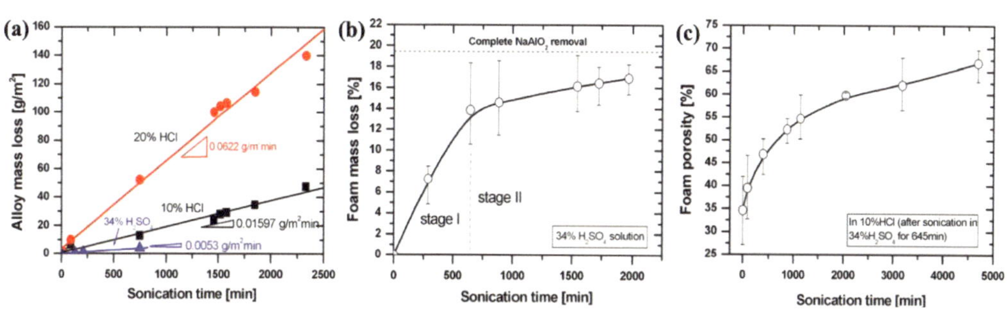

Figure 6. (a) Plot of mass loss vs. time in bulk Ni-Mn-Ga alloy immersed in 10% HCl, 20% HCl, and 34% H_2SO_4; (b) plot of mass loss vs. time for foam with dual-size $NaAlO_2$ powders in 34% H_2SO_4; and (c) plot of foam porosity vs. time for foam with dual-size $NaAlO_2$ powders immersed in 10% HCl after 645 min in 34% H_2SO_4 [35].

to the rapid dissolution stage in **Figure 6b**) and then transferred to the 10% HCl solution. A porosity of 55% was measured after 1140 min in 10% HCl, which was higher than the original $NaAlO_2$ fraction of 45%, indicating the full removal of the $NaAlO_2$ and partial dissolve of the alloy, as shown in **Figure 6c**. Further immersion increased the porosity due to the dissolve of the alloy only.

The morphologies of the fabricated single-pore and dual-pore foams are presented in **Figures 7** and **8**. In foams containing dual pores (**Figure 7b**), the alloy between large pores, which is solid in the single-pore foams (**Figure 7a**), contains small pores, thus producing many small nodes and struts. The three-dimensional architecture of both foams can be seen more vivid in **Figure 8**.

3.2. Ni-Mn-Ga microwires

Metallic microwires can be produced by a variety of methods, that is, melt extraction, Taylor method, melt-spinning, in-rotating water spinning, etc. Details of the fabrication method regarding the fabrication of ferromagnetic wires have been demonstrated by Peng et al. [38]. In recent years, the Taylor-Ulitovsky method was widely used for many metals and alloys. In this process, a metallic ingot is put in a glass tube and melted by induction heating. The glass tube was chosen to have relatively higher melting point than the ingot. Then, the glass tube is softened due to its contact with the molten metal, and it can be drawn. Recently, this method has been modified and applied to create glass-coated Ni-Mn-Ga microwires with equiaxed cross section, as shown in **Figure 9** [39, 40].

In the meantime, melt extraction has been established as a cost-effective and highly efficient method for the production of intrinsically brittle Ni-Mn-Ga alloys on a large scale [13]. The rapid solidification rate during melt extraction is suitable for obtaining refined microstructure, extended elemental solubility, and reduced elemental segregation [41].

A schematic melt extraction facility is displayed in **Figure 10a**. The microwire preparation process consists of the following steps: (1) the Ni-Mn-Ga ingot was inserted into a ceramic

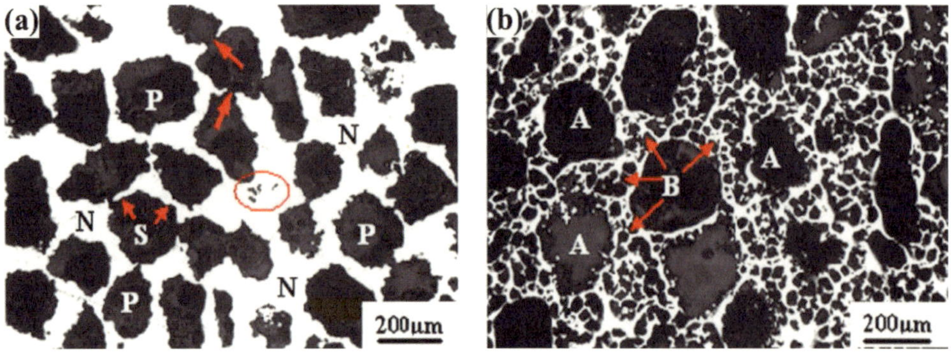

Figure 7. Optical micrographs of polished cross section of Ni-Mn-Ga foams. Metal struts (S) and nodes (N) appear bright, and the pores appear dark. Foams with (a) single and (b) dual-pore size distribution are presented [35].

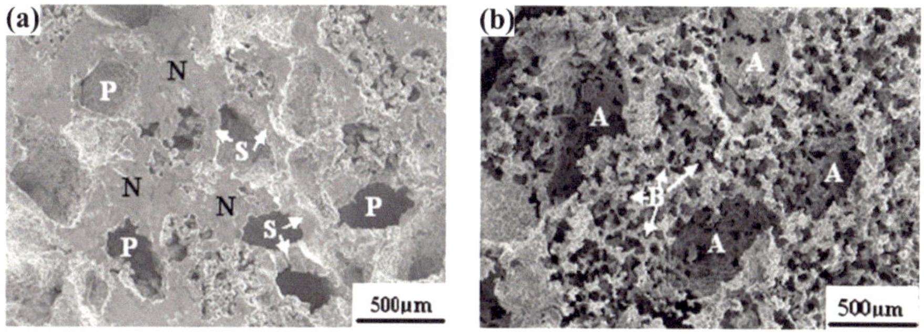

Figure 8. SEM micrographs of cut and etched surface of Ni-Mn-Ga foams showing three-dimensional structure and connectivity of pores. (a) Single- and (b) dual-pore foams are presented [35].

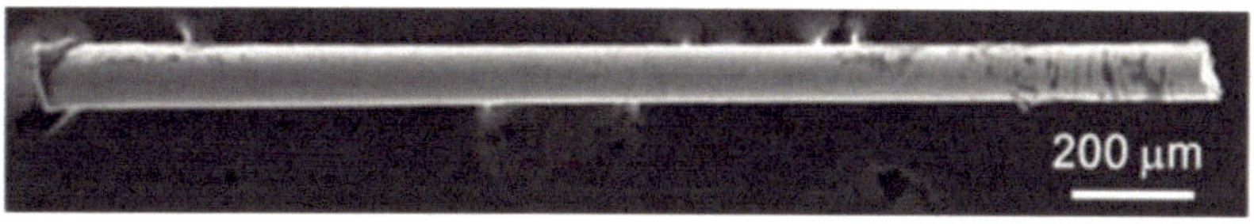

Figure 9. SEM image of representative Ni-Mn-Ga microwires created by the Taylor method [40].

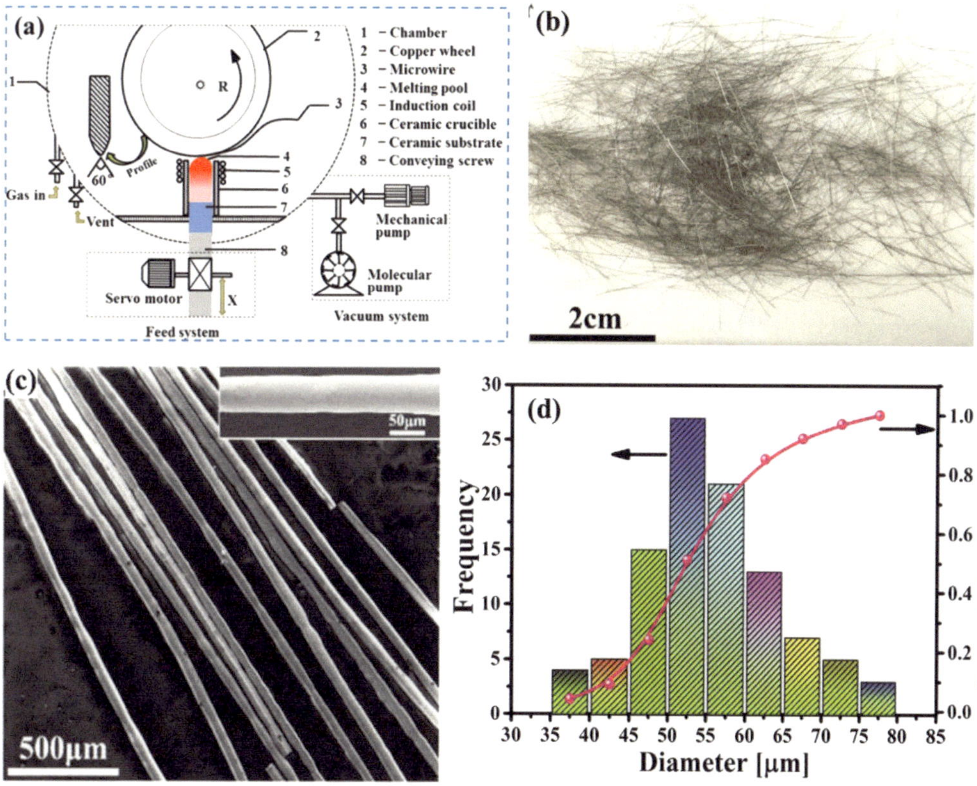

Figure 10. Scheme of the melt extraction process and the obtained Ni-Mn-Ga microwires. (a) Schematic illustration of the melt extraction setup, (b) macroscopic morphology, (c) SEM image, and (d) diameter distribution [13, 18].

crucible and placed into a chamber, (2) the chamber was evacuated to 10^{-3} Pa and then filled with 50 Pa Ar gas, (3) the top part of the ingot was induction heated, thus forming a melting pool in the crucible, (4) the molten phase was driven at a feed rate (V_m) of 40–120 μm/s toward a rotating wheel with wheel rotation velocity (V_w) ranging from 13 to 25 m/s, and (5) the molten alloy was extracted out by the wheel [9]. Typical macroscopic and SEM morphologies of melt-extracted Ni-Mn-Ga microwires are presented in **Figure 10b** and **c**. Microwires with fairly uniform diameter ranging from 45 to 65 μm (**Figure 10d**) were prepared on a large scale with optimized processing parameters: wheel velocity of 23 m/s, feed rate of 60–90 μm/s, and heating power of 20 kW [13].

The microstructural evolution during the melt extraction process, that is, the nucleation behavior of the molten alloy, the unique grain growth behavior, the grain distribution, and the texture, has been systematically studied [13]. As shown in **Figure 11**, microwires with singular nucleation (earlier stage) and dual nucleation (later stage) sites were found at different stages of the preparation: at the beginning of the melt extraction process, the wheel tip temperature is low, and crystallization of the molten alloy nucleates at the wheel tip. Subsequently, the molten alloy nucleates at the two sides of the wheel tip when the wheel tip temperature increased.

Unique grain growth behavior was observed along the radial direction, creating columnar grains growing along the crystal <0 0 1> direction with a fanlike texture in the cross section of the microwire, as shown in **Figure 11**. Finally, at the later stage of the crystallization, the texture evolved into <0 0 1> crystal orientation perpendicular to the flattened surface of the microwire, as shown in **Figure 12**. On this occasion, assuming 5M martensite was formed in the microwires, only variants with *a*-axis [1 0 0] or *c*-axis [0 0 1] perpendicular to the flattened part could exist. The reduced number of twin variants may reduce the incompatibility and thus favor the MFIS.

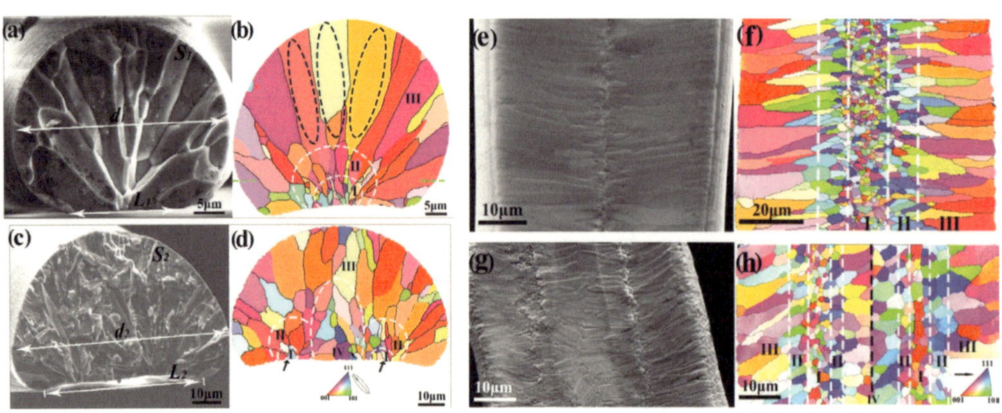

Figure 11. SEM images (a, c, e, g) and EBSD orientation maps (b, d, f, h) of the cross section (a–d) and the longitudinal section (e–h) of the melt-extracted Ni-Mn-Ga microwires [13].

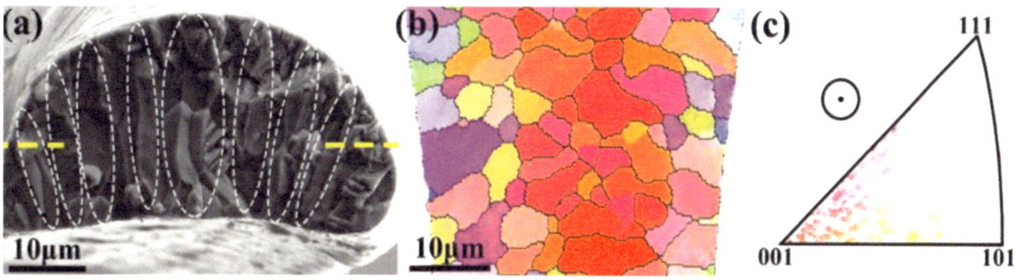

Figure 12. (a) SEM image, (b) EBSD orientation map, and (c) discrete inverse pole figure with respect to (b) of the melt-extracted Ni-Mn-Ga microwires at later stage [13].

4. Martensite transformation of ferromagnetic shape memory alloys

4.1. First-order martensite transformation

4.1.1. Conventional martensite trasnformation

Upon cooling, like in steel, ferromagnetic shape memory alloy undergoes a diffusionless phase transformation from the austenite state to the martensite state, named martensite transformation (MT). Owing to the diffusionless character of the transformation, a slight change of the position of the atoms in cubic austenite phase during cooling leads to a tetragonal distortion of the unit cell.

As was mentioned in Section 2.2, Heusler alloys possess two important phases: the cubic austenite phase and the martensite phase (with different lattice structures). The MT process can be illustrated based on **Figure 13** [17]: upon cooling from an austenite state, the sample starts to form martensite at the martensite start temperature, M_s. The sample is fully transformed to martensite state below the martensite finish temperature, M_f. During the reverse transformation when the sample heats from a fully martensite state, the austenite starts to form above the austenite start temperature, A_s, and is completely transformed to austenite above the austenite finish temperature, A_f.

As shown in the DSC curve of **Figure 13** (upper), the transition from the austenite to martensite is an exothermic reaction, which means that heat is released for this process and the enthalpy change of the system is negative. On the other hand, the reverse transformation from martensite to austenite is an endothermic reaction with a positive enthalpy change. Besides, during DSC measurements, the temperatures at maximum endothermic and exothermic heat flow are defined as A_p and M_p, respectively. While as shown in the magnetization-temperature (*M-T*) curves of **Figure 13** (lower), the A_p and M_p are derived from the peak values of the first derivative plot of the curves during heating and cooling, respectively (shown in the inset). Martensite transformation is a first-order phase transformation (FOT). A FOT is usually accompanied by an inevitable transformation hysteresis, which can be interpreted by $|A_p - M_p|$ or $(A_s - M_s + A_f - M_f)/2$.

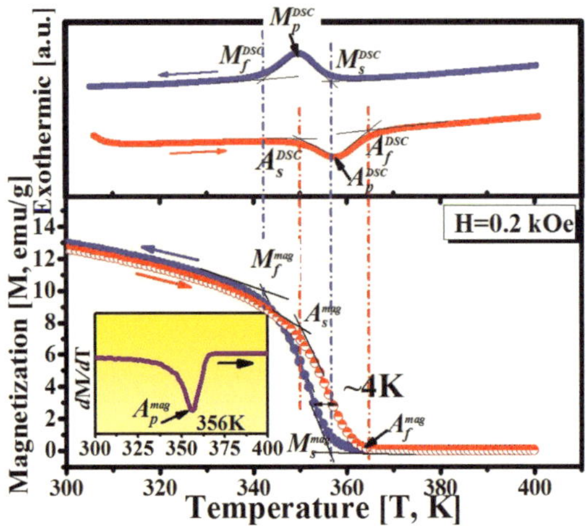

Figure 13. Heating and cooling of DSC curves (upper) and *M-T* curves at low magnetic field of ferromagnetic shape memory alloys [17].

The MT temperatures are very sensitive to the composition for the alloys. For Ni-Mn-Ga and other Ni-Mn-X alloys, the composition dependence of the MT temperatures has been demonstrated in Section 2.1 of the present chapter.

4.1.2. Premartensite transformation

Premartensite transformation (PMT), a weak FOT of the austenite into a micromodulated premartensite phase prior to the martensite transformation itself, has been observed in Ni-Mn-Ga alloys with T_c well above the MT temperature, that is, mainly in Ni-Mn-Ga alloys from Group I (**Table 1**) with near-stoichiometric composition [31, 42]. A PMT at ~260 K has been found in Ni_2MnGa alloys prior to its MT [43], with the austenite phase transformed to an orthogonal three-layer modulated structure, which is similar as the 5M and 7M modulated structure [44].

Existence of unusual physical property change has been found around the PMT temperature during cooling. Firstly, before the PMT, the samples started to exhibit a drastic increase of elastic modulus [32]. Furthermore, the internal friction is increasing from zero to a maximum in the premartensite phase and decreases again in martensite [31]. This sudden soft mode phonon freezing can only be observed in the premartensite phase before the MT; thus, this transformation is defined as the PMT.

4.1.3. Intermartensite transformation

Intermartensite transformation (IMT), a structural transformation from one martensite to another, has been found in high-temperature Ni-Mn-Ga alloys, that is, mainly in Ni-Mn-Ga alloys from Group III (**Table 1**). Upon cooling in these samples, the austenite phase changes to a 5M then 7M and then non-modulated phase or to 7M and then non-modulated phase or directly from austenite to non-modulated phase [32]. While upon heating, only a phase transformation from the non-modulated martensite to austenite was observed [45]. On the other hand, the IMT can also be induced with increasing stress following the same transformation routes as is induced with decreasing temperature [32].

4.2. Second-order magnetic transformation

The increased distance between the Mn atoms in the Ni-Mn-Ga $L2_1$ structure changes the Mn-Mn exchange interaction from antiferromagnetic of pure Mn to ferromagnetic [33]. Therefore, a magnetic transformation is taking place at the Curie temperature (T_c) from a ferromagnetic phase to a paramagnetic phase upon heating. The magnetic transformation is a second-order transformation (SOT) with no latent heat associated with the phase transformation and can be easily detected from the M-T curves with a drastic falling of the magnetization. The T_c of the Heusler alloys is also influenced by the composition and, thus, can be tuned. In our previous work, the MT temperature was increased, and the T_c was lowered after Cu doping in Ni-Mn-Ga alloy, leading to an overlap of the martensite and magnetic transformation, that is, magneto-structural coupling, in the microwire. The magneto-structural coupling enhanced the MCE of the alloy [17].

5. Properties and application of ferromagnetic shape memory alloys

5.1. Magnetic-field-induced strain (MFIS)

5.1.1. Overview of MFIS

Magnetic-field-induced strain (MFIS) comes from twin boundary motion under the application of a magnetic field, which is driven by the magnetostress produced by high magnetocrystalline anisotropy of the Ni-Mn-Ga alloy and its low twining stress [35], as schematically illustrated in **Figure 14** [46]. The MFIS property is useful for actuation and sensing purposes [47]. MFIS was firstly reported in 1996 by Ullakko et al. [5]. A strain of 0.19% under a magnetic field of 0.43 T was obtained from a Ni_2MnGa sample at 265 K. From then on, the interest of the MFIS in FSMAs grew rapidly all over the world. Besides, many different compositions, such as Ni-Fe-Ga [48], Ni-Mn-In [49], Co-Ni-Al [50], Co [51], Fe [52], and rare earth [53]-doped Ni-Mn-Ga alloys, have also been investigated.

So far, large MFIS (~1–10%) has been achieved only for Ni-Mn-Ga single crystals [6, 7]. The most representative one was reported in 2000, when Murray et al. [6] achieved a 6% MFIS in a 5M single-crystalline Ni-Mn-Ga alloy at room temperature under a magnetic field of 0.62 T. During the experiment, different stresses were applied to restore the MFIS and to measure the magnetostress. After that, in 2002, Sozinov et al. [54] published a giant MFIS of about 9.5% in 7M Ni-Mn-Ga single crystals at ambient temperature in a magnetic field of less than 1 T. However, the fabrication of single crystals is difficult because of sever segregation, low growth speed, and high cost. On the other hand, polycrystalline Ni-Mn-Ga alloys may be produced with much lower cost, but their MFIS is vanishingly small (<0.01%) because of the low mobility of twin boundaries constrained by grain boundaries [55, 56]. Many works have been

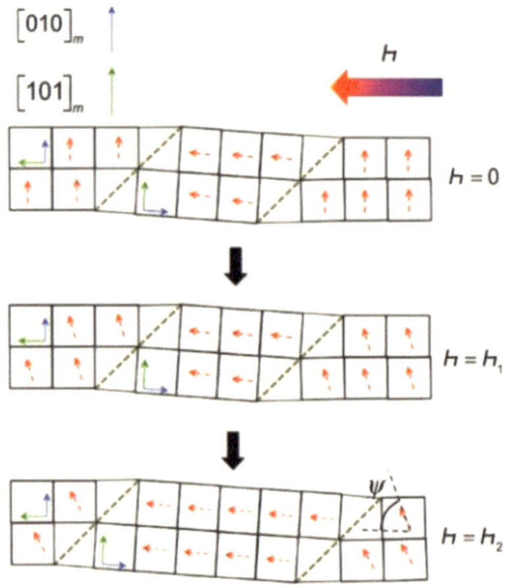

Figure 14. Schematic illustration of rotation of magnetic moments and twin boundary motion in tetragonal FSMAs under magnetic fields [46].

carried out to enhance their MFIS up to 1% by introducing strong textures and subsequent training in coarse-grained polycrystalline Ni-Mn-Ga alloys [57–60]. Recently, considering the hindering effect of the grain boundaries on the twin boundary motion, approaches regarding reduction of grain boundaries, that is, by producing a porous material (foam) or reducing sample size, have been carried out, which will be discussed in detail in Sections 5.1.2 and 5.1.3.

5.1.2. MFIS in Ni-Mn-Ga foams

As mentioned earlier, fine-grained polycrystalline Ni-Mn-Ga alloys are easier to fabricate but with vanishingly small strain due to the constraints provided by the grain boundaries. Introducing porosity in Ni-Mn-Ga alloys not only reduces the constraints imposed by grain boundaries but also maintains the ease of processing associated with casting polycrystalline Ni-Mn-Ga. After certain grain growth heat treatment, the twins can span between the pores, as shown in **Figure 15** [10]. As a result, the twin boundaries can move as freely as in single-crystalline bulk material within the grains.

Research work regarding the MFIS in Ni-Mn-Ga foams was firstly reported by the research group of Müllner et al. [9]. With 76% open porosity, the foam displayed a fully reversible MFIS as large as 0.12% with excellent stability over 25 million magnetomechanical cycles. Thereafter, the same group further increased the MFIS to 2.0–8.7% in dual-pore Ni-Mn-Ga foams (**Figures 7b** and **8b**) with 62% porosity after thermo-magneto-mechanical cycling training [10]. These obtained strains are much larger than those of any polycrystalline and comparable to those of single crystals. Different from bulk single or polycrystalline alloy, these open-porosity foams allow fluid flow, making them potentially useful as micro-pumps and magnetocaloric materials [10].

http://dx.doi.org/10.5772/intechopen.69702

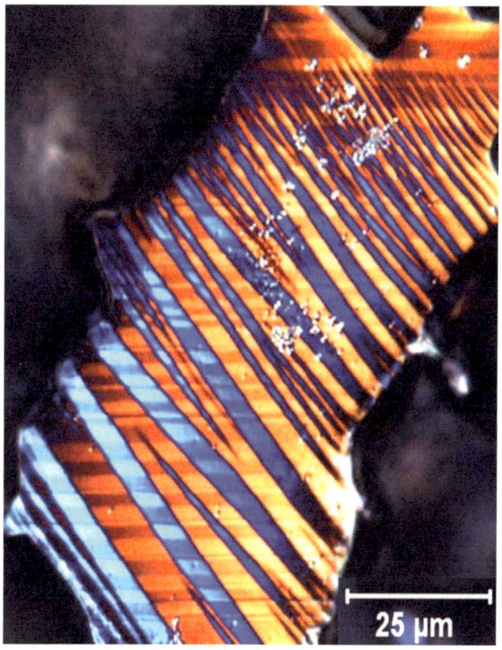

Figure 15. Optical micrograph of twins in Ni-Mn-Ga foam, extending entirely from pore to pore (black) [10].

5.1.3. MFIS in Ni-Mn-Ga microwires

Constraint of twin boundary caused by grain boundary is three dimensional in bulk alloys and two dimensional in thin films. It can be further reduced to one dimensional in microwires. Furthermore, by reducing sample size, small-sized Ni-Mn-Ga alloys show low inertia [61], low eddy current loss at high frequency [62], high magnetocrystalline anisotropy, and work output [63]. Recently, oligocrystalline microwires with bamboo or near-bamboo structures have attracted much interests owning to a less constrained environment and more free surface [64–66], which makes them the closest approximation to single crystals. The micrographs of Ni-Mn-Ga and Cu-Al-Ni oligocrystalline microwires are presented in **Figures 16** and **17**, respectively. Martensite plates spanning across the wire diameter can be observed [65, 66].

However, because of the lack of textures, only a subset of grains was prone to favorably oriented to show a detectable MFIS. Approximately 1% MFIS was found in a not-constrained and randomly textured Ni-Mn-Ga without bias stress by magnetizing the microwire parallel and perpendicular to the wire axis up to 2 T [11]. Recently, a 1 T rotating magnetic field caused the Ni-Mn-Ga microwire to bend to a curvature corresponding to a surface strain of 1.5% [40]. Mechanical or the combined thermo-magneto-mechanical training in martensite state may effectively lower the twining stress, to form preferential oriented variants [9, 58], and, thus, may in favor of achieving a high MFIS in oligocrystalline microwires. The realization of large MFIS in one-dimensional microwires may provide a wide prospect in the application of micro-actuators and sensors.

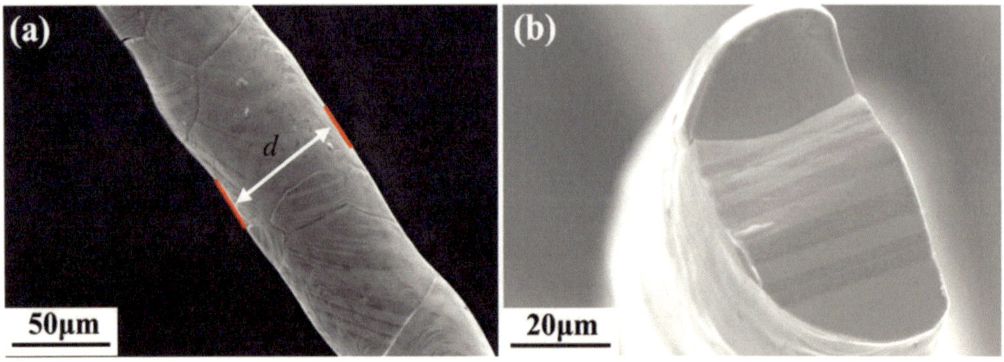

Figure 16. SEM micrographs of (a) free surface and (b) cross section of Ni-Mn-Ga oligocrystalline microwires prepared by melt extraction and subsequent heat treatment [66].

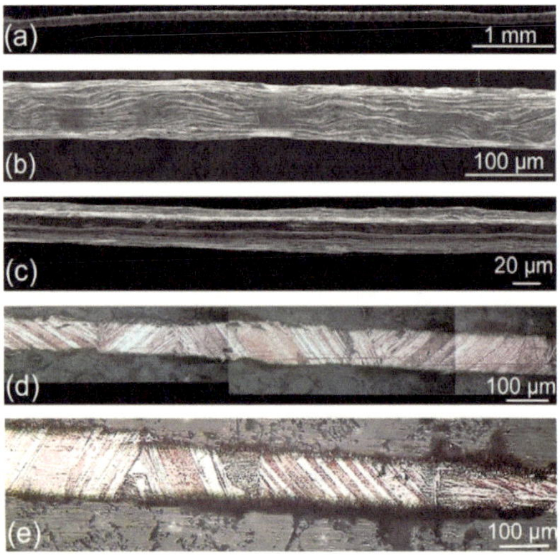

Figure 17. (a–c) Micrographs of Cu-Al-Ni oligocrystalline microwires prepared by Taylor method and subsequent heat treatment. (d and e) Montaged optical micrographs of the longitudinal section of the wire [65].

5.2. Magnetocaloric effect (MCE)

5.2.1. Overview of MCE

Magnetic refrigeration, based on the magnetocaloric effect (MCE), has attracted interests as a potential alternative to well-established compression-evaporation technique for room temperature refrigeration because of the compactness, high efficiency, and environmental friendship. To characterize a MCE, the adiabatic temperature change (ΔT_{ad}), the magnetic entropy change (ΔS_M), and the relative cooling power (*RCP*) should be measured. The most recently researched ambient magnetic refrigeration materials (MCM) mainly include $La(Fe,Si)_{13}$ based [67–70] and $Gd_5(Si,Ge)_4$ based [67, 71] alloys (contain rare-earth elements) and MnAs based [67, 72–74],

MnFe(P,X) (X = As, Ge, Si) [75, 76], Fe-Rh [77], NiMn based [78–82] alloys (rare-earth free). These types of materials undergo a first-order magnetic phase transition (FOMT), exhibit large hystereses, and show a large value of ΔS_M. For example, with a magnetic field change $\Delta H = 5$ T, $\Delta S_M = 35–50$ J/kg K for MnAs-based [73], $Gd_5(Si,Ge)_4$-based [67], and $La(Fe,Si)_{13}$-based [83] alloys and $\Delta S_M = 35–40$ J/kg K for Ni-Mn-In-Co [84] and Ni-Mn-Sn [85] alloys were obtained. For some MCE, the low working temperature span (ΔT_{FWHM}) and high thermal and magnetic hysteresis loss [86, 87] limit their applications. MCM with reduced dimensions (particle, microwire, film, or foam) [88, 89] have been proposed to broaden the ΔT_{FWHM} by preparing the alloy with gradient composition distribution state. In addition, deduction in dimensions is an effect way to decrease the hysteresis loss by reducing internal stress and the constraints between grains attributed to the high surface-area-to-volume ratio [19].

5.2.2. MCE in Ni-Mn-based alloys

Very recently, we reported magnetocaloric effects of high-content Fe-doped $Ni_{44.9}Fe_{4.3}Mn_{38.3}Sn_{12.5}$ polycrystalline microwires prepared by a melt extraction technique [18], as displayed in **Figure 18**. Under a magnetic field of 20 kOe, the ΔS_m peak value, related to the first-order martensite transformation (FOMT) of the present $Ni_{44.9}Fe_{4.3}Mn_{38.3}Sn_{12.5}$ microwires, reaches 3.0 J/kg K, which is comparable to the $Ni_{44}Fe_6Mn_{40}Sn_{10}$ ribbons (1.8 J/kg K under 20 kOe) [90] and $Ni_{51.6}Mn_{32.9}Sn_{15.5}$ films (1.5 J/kg K under 10 kOe) [91]. The second-order martensite transformation (SOMT) related to ΔS_m of −3.7 J/kg K and ΔT_{FWHM} of ~85 K under 5 T is comparable to that of $Ni_{48.8}Mn_{26.7}Ga_{20.8}Cu_{3.7}$ microwires (−8.3 J/kg K with ΔT_{FWHM} of 13 K under 5 T) [19] and $Ni_{53.5}Mn_{23.8}Ga_{22.7}$ films (−8.5 J/kg K with ΔT_{FWHM} of 17 K under 6 T) [92]. The studied $Ni_{44.9}Fe_{4.3}Mn_{38.3}Sn_{12.5}$ microwires, exhibiting large surface-area-to-volume ratio, giant MCE property, and low cost, may act as potential magnetic refrigerants.

As shown in **Figure 19a**, the ΔS_m of $Ni_{48}Mn_{26}Ga_{19.5}Fe_{6.5}$ microwires [19] with a diameter of ~50 μm shifted from positive to negative to the applied field is higher than 0.5 T. These interesting positive-to-negative ΔS_m transition behaviors have also been found in $Ni_{50.1}Mn_{20.7}Ga_{29.6}$ single crystal at 0.8 T [93] and $Ni_{54.5}Fe_1Mn_{20}Ga_{24.5}$ polycrystalline alloy at 0.5 T [94]. The maximum negative ΔS_m is ~4.7 J/kg K under 5 T corresponding to magnetic-field-induced structural transition from martensite to austenite phase. At 20 kOe, ΔS_m in the microwires is 1.71 J/kg K, which is comparable to Ni-Fe-Mn-Ga alloy (2.1 J/kg K) [94]. Moreover, the value of ΔS_m observed at magnetic transition under 3 T is 2.05 J/kg K in Fe-doped $Ni_{54.5}Fe_1Mn_{20}Ga_{24.5}$ microwires. Compared to the un-doped $Ni_{50.95}Mn_{25.45}Ga_{23.6}$ microwires, the ΔS_m is about three times higher under the same magnetic field [95].

It can be seen from **Figure 19b** that the annealed $Ni_{48.8}Mn_{26.7}Ga_{20.8}Cu_{3.7}$ microwires [87] exhibited magneto-structural coupling and wide martensitic transformation temperature range, which contribute to a ΔS_m of 8.3 J/kg K with a wide ΔT_{FWHM} of 13 K under a magnetic field of 5 T. The obtained RC in Ni-Mn-Ga-Cu microwire (78.0 J/kg) is comparable with those of Ni-Mn-based alloys (70–115 J/kg) [96] and superior to those of Ni-Mn-Ga-Cu bulk alloys (72–75 J/kg) [97–99]. On the other hand, when compared to Gd [100] or $LaFe_{13-x}Si_x$ [101] alloys, the Ni-Mn-Ga-Cu microwires are rare-earth free and thus cost-effective, which helps for the practical applications.

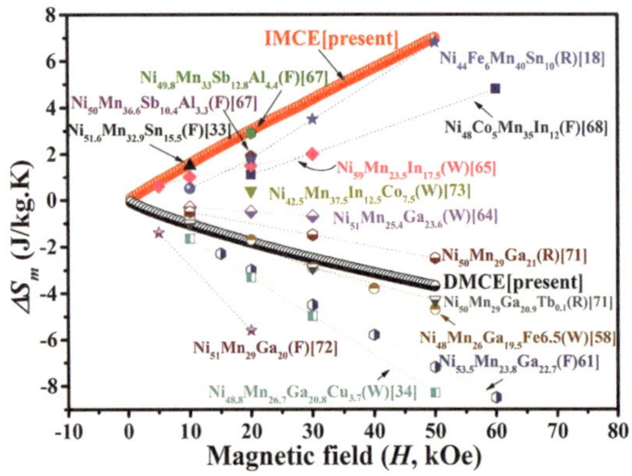

Figure 18. The ΔS_m as a function of external magnetic field in the present $Ni_{44.9}Fe_{4.3}Mn_{38.3}Sn_{12.5}$ microwires and some other Ni-Mn-Z-based (Z = Sn, Sb, In, Ga) Heusler compounds in the forms of ribbons (R), thin films (F), and microwires (W) [18].

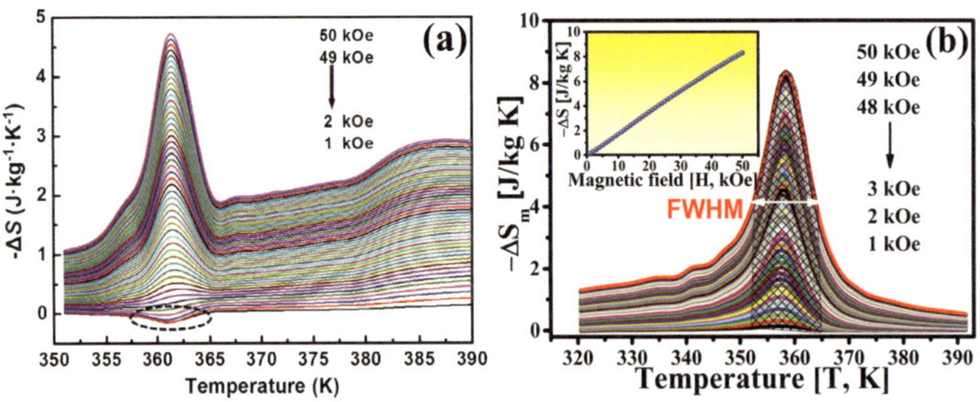

Figure 19. Entropy changes as a function of temperature for (a) $Ni_{48}Mn_{26}Ga_{19.5}Fe_{6.5}$ [19] and (b) $Ni_{48.8}Mn_{26.7}Ga_{20.8}Cu_{3.7}$ [17] microwires under different magnetic fields.

5.3. Shape memory and superelasticity

In Sections 5.1 and 5.2, Ni-Mn-Ga alloy has been demonstrated to exhibit high MFIS and excellent MCE properties both driven by external magnetic field. Actually, due to the thermoelastic MT, Ni-Mn-Ga alloys also show well-pronounced thermal field-induced superelasticity (SE, stress-induced MT at austenite state and recovered upon unloading) [45], one-way shape memory effect (OWSME, deformed under stress at martensitic state and recovered during heating), and two-way shape memory effect (TWSME, shape change between martensite and austenite states continuously upon heating and cooling) [8, 9], which is similar to traditional SMAs, such as Ni-Ti.

So far, OWSME strain up to ~6.1% [102] and SE ~6% [45] has been achieved in Ni-Mn-Ga single crystals under compression mode. TWSME induced by training has reached 9% in

http://dx.doi.org/10.5772/intechopen.69702

single-crystalline Ni-Mn-Ga alloy [8] under tension mode. On the other hand, related reports for SME and SE properties for bulk polycrystalline Ni-Mn-Ga alloys are relatively seldom and lower than those of single crystals [102, 103]. Besides, due to the intrinsic brittleness of the alloy, previous reports are mostly focused on the compression processes than on the tension mode.

5.3.1. Shape memory effect and superelasticity in Ni-Mn-Ga microwires

Due to the transcrystalline fracture tendency of polycrystalline Ni-Mn-Ga alloys, efforts have been made to enhance their ductility. Microstructure refinement and sample size reduction have been proven to lower the brittleness [6, 9]. Recently, small-sized Ni-Mn-Ga microwires have been made by rapid solidification methods involving Taylor method and melt extraction technique to investigate the SME and SE properties [12, 14, 39, 40]. A large reversible SE strain of 10.9% has been reported in polycrystalline Ni-Mn-Ga glass-coated wires tested in tension mode [39]. The Taylor method as applied to Ni-Mn-Ga microwire fabrication allows production of wires with a uniform circular cross section while with lower efficiency compared with the melt extraction technique. In this section, the SME and SE in melt-extracted microwires are displayed based on our previous work.

Ni-Mn-Ga microwires prepared by melt extraction technique exhibited a higher reversible SE strain and recoverable SME strain compared with bulk parent alloys [12]. **Figure 20** displays the typical recoverable SME curve of as-extracted Ni-Mn-Ga microwire, and A–C demonstrates the SME strain regarding twin boundary motion after unloading (~1.5%). Unlike single crystals, the twin boundary motion process did not show a stress plateau upon loading (A–B) due to the refinement of the grains.

Owing to the rapid solidification process during fabrication, high internal stresses, reduced degree of atomic order, and other defects may affect the SME and SE properties in the as-extracted microwires. A stepwise chemical ordering annealing was carried out, and the effect of annealing on the SE behavior was investigated in our previous work [14]. SE comparison before and after annealing is shown in **Figure 21**. Annealing decreases the stress-induced MT (SIM) stress and the hysteresis and improves the reversibility during superelastic cycling.

Furthermore, polycrystalline Ni-Mn-Ga microwires exhibit higher-temperature dependences compared with Ni-Mn-Ga single crystals and conventional superelastic alloys (**Figure 22**), which are considered to be related to the small grains achieved by melt extraction. Besides, the temperature dependences of the microwires are lowered after annealing (inset II), that is, the slope ($d\sigma_{As}/dT$) of the annealed microwire, 15.6 MPa/K, is smaller than that of as-extracted one, 17.5 MPa/K, revealing an easier reverse transformation process in the annealed microwire.

Above all, melt-extracted Ni-Mn-Ga microwires after annealing exhibit lower SE stresses, lower-temperature dependences, and higher SE reversibility. Besides, the preparation is convenient and efficient. Given these properties, Ni-Mn-Ga microwire is expected to be used for practical applications, such as superelastic materials, micro-actuator materials, and microsensor materials in various fields.

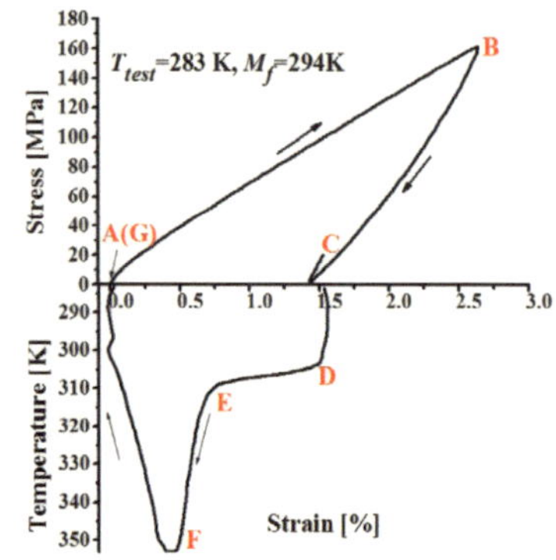

Figure 20. Tensile stress-strain curves of as-extracted Ni-Mn-Ga microwires at martensite state.

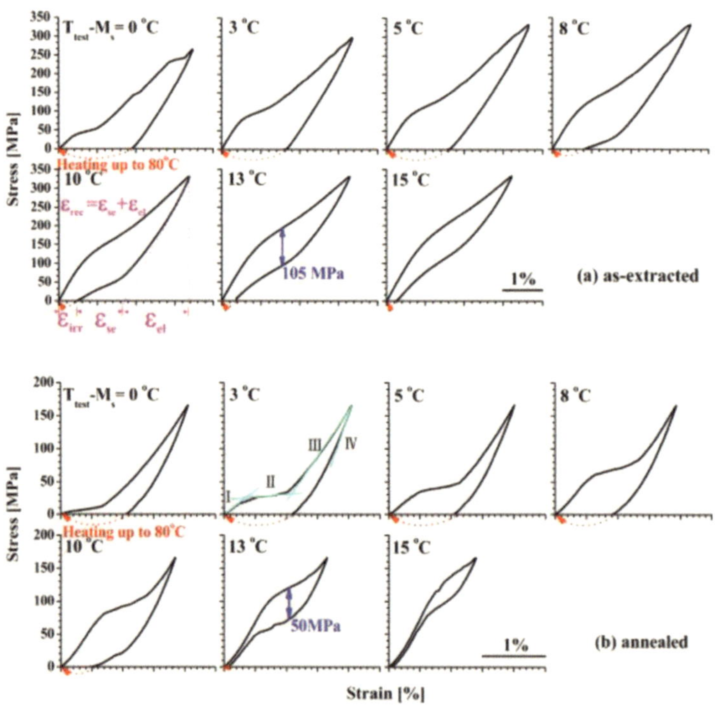

Figure 21. Tensile stress-strain curves of (a) as-extracted and (b) annealed Ni-Mn-Ga microwires at temperature levels (T_{test} − M_s) approximately from 0 to 15°C at austenite state (equilibrate at 80°C and then cool down to T_{test}). ε_{irr}, ε_{se}, and ε_{el} stand for the irreversible strain (recovered upon heating), strain recovery upon reverse transformation, and elastic recovery, respectively [14].

5.3.2. Shape memory effect and superelasticity in Ni-Mn-Ga-Fe microwires

The ductility of Ni-Mn-Ga alloys can be enhanced by the fourth element doping [104–106]. Among various doping elements, Fe has attracted many attentions [107, 108]. In our previous work, SME and SE behaviors of $Ni_{50}Mn_{25}Ga_{25-x}Fe_x$ (x = 1–6%) microwires (diameter ~30–40 μm)

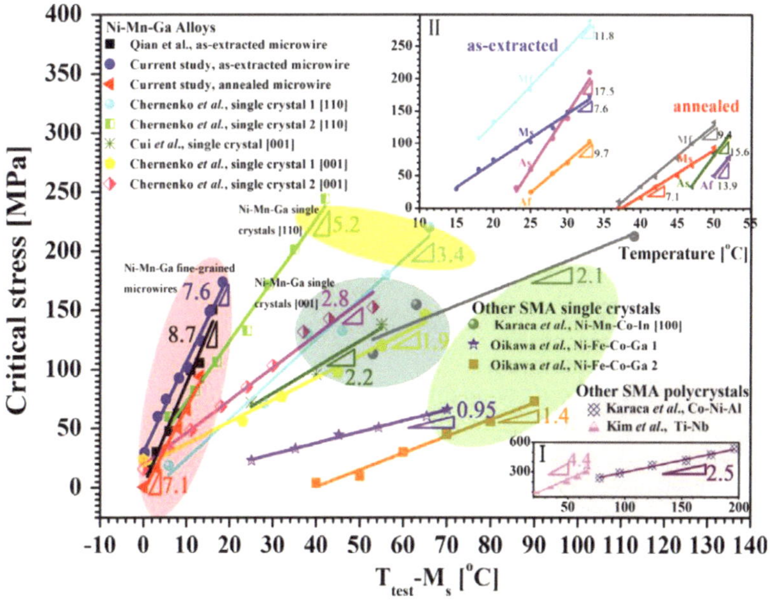

Figure 22. SIM critical stress (σ_{Ms}) vs. temperature plots of the Ni-Mn-Ga microwires and other alloys [14].

were investigated [16]. OWSME and TWSME curves of the $Ni_{49.7}Mn_{25}Ga_{19.8}Fe_{5.5}$ microwires are shown in **Figure 23a** and **b**, respectively. An OWSME strain of ~1.0% was induced by stress at its martensite state under tension mode in the microwire and completely recovered upon heating to its austenite state (**Figure 23a**). With respect to the TWSME, the application in sensors is generally used under certain applied stress in thermal cycle. Different external stresses were applied during the thermal cycling; fully recoverable strain was obtained in the

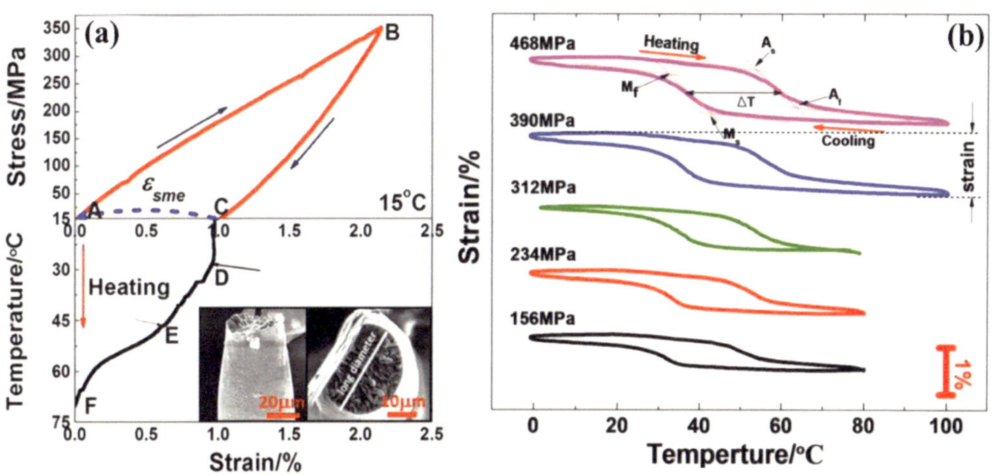

Figure 23. Tensile stress-strain curves of $Ni_{49.7}Mn_{25}Ga_{19.8}Fe_{5.5}$ microwires. (a) OWSME and (b) TWSME [16].

$Ni_{49.7}Mn_{25}Ga_{19.8}Fe_{5.5}$ microwire (**Figure 23b**). The TWSME strain of the microwire increased from 0.84% at 156 MPa to 1.504% at 468 MPa.

The SE behavior of the $Ni_{50}Mn_{25}Ga_{25-x}Fe_x$ (x = 4.5) was studied due to its favorable MT temperature near RT. The SE tensile stress-strain curves obtained at various temperatures in $Ni_{50}Mn_{25}Ga_{25-x}Fe_x$ (x = 4.5) are demonstrated in **Figure 24**.

The SE of $Ni_{50}Mn_{25}Ga_{25-x}Fe_x$ (x = 4.5) microwires shows similar behavior as that of Ni-Mn-Ga microwires [9]. The maximum strain recovery rates achieved in $Ni_{50}Mn_{25}Ga_{25-x}Fe_x$ (x = 4.5) are 94 and 90%, respectively, which was higher than that in as-extracted Ni-Mn-Ga microwires while lower than Ni-Mn-Ga microwires after chemical ordering annealing.

5.4. Application of ferromagnetic shape memory alloys

Ni-Mn-Ga alloys may act as actuator by making the use of the large MFIS and magnetic shape memory effect (MSM) or as sensor related to the dependence of the magnetization change under an external compressive stress. For MFIS/MSM, the working frequency is high because they are driven by external magnetic fields. This character is superior to the traditional shape memory alloys, such as Ni-Ti, which are driven by temperature change and thus only work at much smaller frequency. Some typical examples of the actuators created based on the ferromagnetic shape memory alloys are shown in **Figures 25–27**.

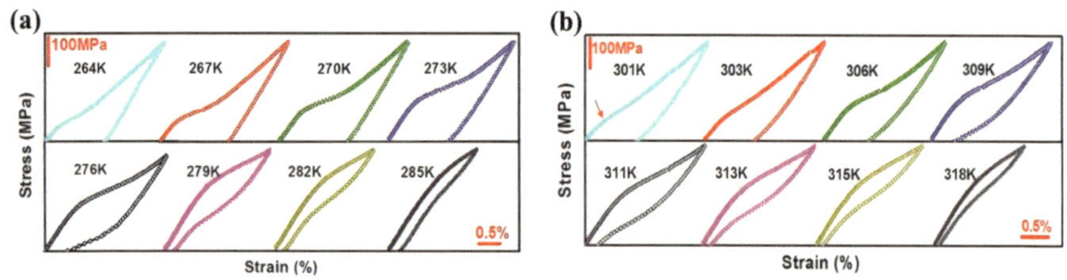

Figure 24. Tensile stress-strain curves obtained at various temperatures in $Ni_{50}Mn_{25}Ga_{25-x}Fe_x$ microwires. (a) x = 4 and (b) x = 5 [15].

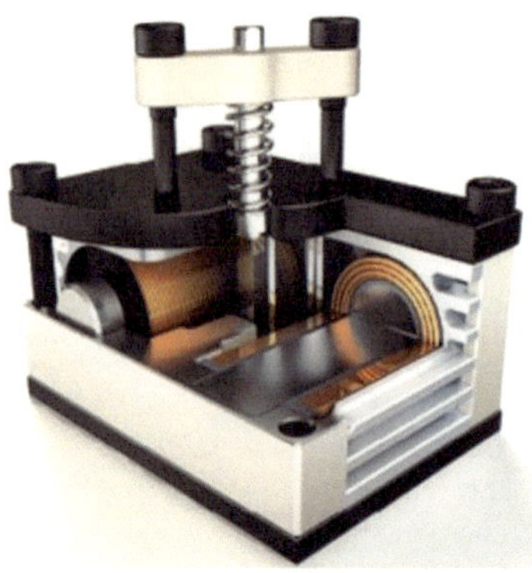

Figure 25. MSM-spring actuator (MAGNETOSHAPE® by ETO MAGNETIC GmbH) [109].

http://dx.doi.org/10.5772/intechopen.69702

Figure 25 demonstrates the most straightforward MSM-spring actuator: the elongation is obtained by a magnetic field perpendicular to the motion, while contraction is obtained thanks to the elastic force of a spring. The magnetic field induces a magnetostress in the MSM alloy. In the spring actuator, such a magnetic force must work always against the elastic force of the spring, which is bigger at bigger strain [109].

Figure 26a presents another important MSM actuator: push-push or multistable actuator [109]. It is composed of two MSM units which arranged antagonistically. In this case, one element acts as a load for another one. The movement in both directions can be controlled magnetically. There is the moving rod in yellow on the top. **Figure 26b** shows an application of a push-push actuator: the MSM device is used to move a mirror on the top and redirect an optical signal.

For FSMA thin films, a novel actuation mechanism has been developed, which makes use of both the ferromagnetic transition and the martensitic transformation. The mechanism is illustrated in **Figure 27** for a Ni-Mn-Ga bending actuator placed in the inhomogeneous magnetic field of a miniature permanent magnet.

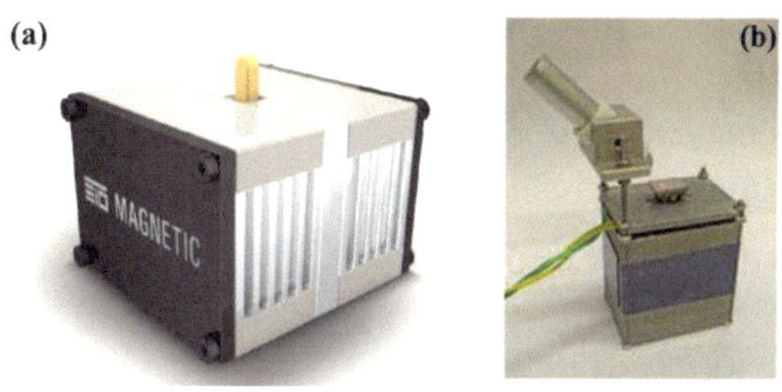

Figure 26. (a) Example of MSM push-push actuator (MAGNETOSHAPE® by ETO MAGNETIC GmbH). (b) Optical switch actuated by a push-push actuator (Lappeenranta University of Technology, Finland) [109].

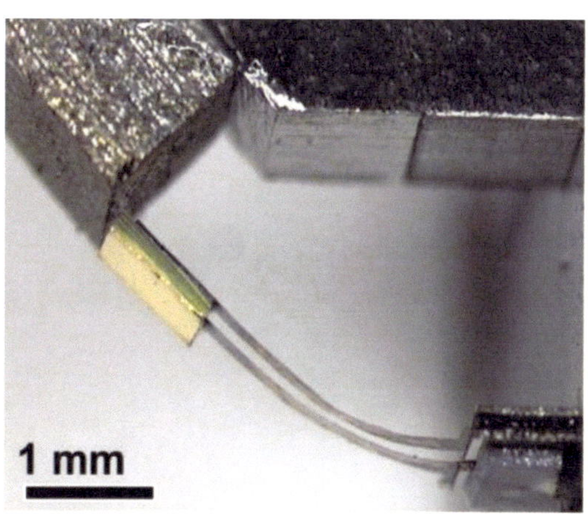

Figure 27. Prototype of a FSMA micro-actuator for control of a micro-mirror [110].

Depending on the temperature of the micro-actuator, either magnetic or shape recovery forces occur in opposite directions, while the corresponding biasing forces remain small. Thus, an almost perfect antagonism can be realized in a single component part. By applying an alternating electrical current, a periodic oscillation of the beam can be excited. This motion can be used to control the deflection of a micro-mirror attached to the front end of the actuator. For instance, a micro-scanner prototype has been developed based on the actuation mechanism (see **Figure 27**). The overall dimensions are 7 mm × 2 mm × 5 mm.

The micro-pump shown in **Figure 28**, developed in Peter Müllner's lab at Boise State University, may deliver sub-microliter volumes of dugs directly to specific regions of the brain. The micro-pump was small and robust and can be placed on a head stage on a rat so that drugs can be delivered and brain activity can be monitored while the rat is moving about [111].

Figure 29 shows a prototype of an energy harvester realized by former Adaptamat [109]. A repeated application of tension and compression force deforms the MSM material and induces a voltage in the coil. Such a voltage can be used to supply energy to a small load. Theoretically, the maximum energy per volume unit that can be extracted from an MSM alloy is equal to the work output, that is, about 150 kJ/m^3.

In Ni-Mn-Ga alloys, the first-order martensite transformation (FOMT) and second-order magnetic transition (SOMT) produce traditional MCE [112], as displayed in **Figure 30a**. On the other hand, in Ni-Mn-Z-based (Z = Sn, In, Sb) alloys, both direct and inverse MCE (see **Figure 30b**) may be created. The inverse MCE originates from a metamagnetic structural

Figure 28. Micro-pump created based on Ni-Mn-Ga alloys [111].

Figure 29. Energy harvester based on MSM alloys developed by Adaptamat [109].

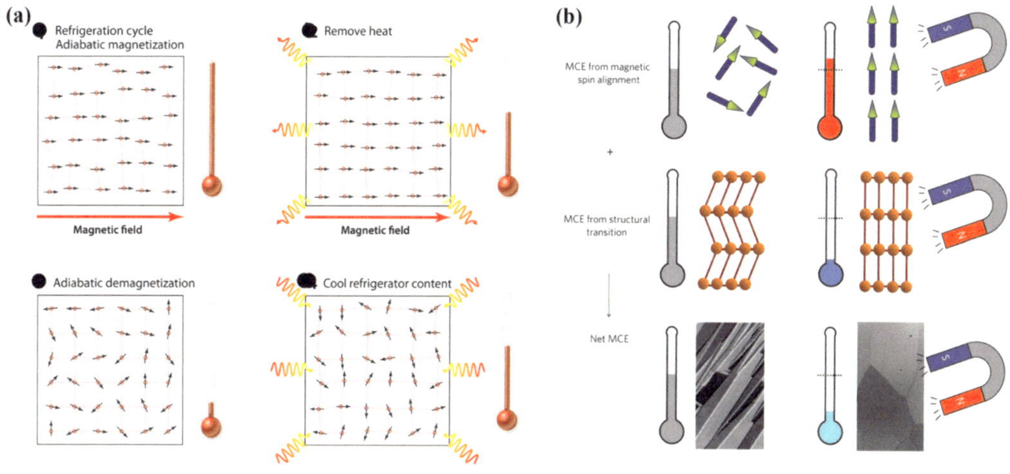

Figure 30. Schematic illustration of (a) traditional magnetocaloric effect [112] and (b) inverse magnetocaloric effect [21].

transition from the paramagnetic/antiferromagnetic martensite to ferromagnetic austenite under a bias magnetic field [21], while the direct MCE is attributed to the magnetic transition of the austenite phase around its Curie point [113]. The first-order martensite transformation (FOMT) is responsible for such inverse MCE, which usually exhibits a huge ΔS_m and a large adiabatic temperature change ΔT_{ad} but rather low ΔT_{FWHM} and high hysteresis loss [86]. Ni-Mn-In-Co alloys may also produce giant stress output through magnetic-field-induced martensite transformation [20].

6. Conclusions remarks

The martensite transformation temperature of FSMAs is sensitive to the composition, while the magnetic transition temperature is less sensitive to the composition. Some empirical formula has been summarized to build the relationship between the transformation temperatures and compositions. In addition, the martensite and austenite crystal structures as well as the magnetic properties of these alloys have been extensively investigated.

Single-crystalline FSMAs, such as Ni-Mn-Ga alloys, have been widely studied because the high MFIS is usually generated in the single-crystalline alloys since low resistant to the twin boundary motion. The compositional segregation during the growth of single-crystalline alloys has to be carefully controlled in order to fabricate the alloy ingots with repeatable MFIS. On the other hand, polycrystalline alloys may be fabricated by low-cost methods, such as casting. The polycrystalline Ni-Mn-Ga foams produced by replication casting may generate MFIS as high as 8.7% after suitable training.

Polycrystalline microwire may be synthesized on a large scale by melt extraction, which exhibits pronounced properties, such as MFIS, magnetic entropy change, and superelasticity. Small-sized materials, such as powders, microwires, ribbons, and films, exhibit giant-specific surface area, that is, surface-area-to-volume ratio, which is responsible for the reduced constraints to the twin boundary motion, enhanced heat exchange efficiency, and improved magnetic refrigeration hysteresis loss. The underlying mechanisms between the material size and MCE properties need to be further studied.

The martensite transformation in FSMAs may be induced by external heat change, similar to conventional shape memory alloys, such as Ni-Ti, as well as by external magnetic field. As the operation frequency of an external magnetic field can be much higher than a heat field, FSMAs can work at much higher frequencies than conventional Ni-Ti alloys. The high MFIS produced in single-crystalline Ni-Mn-Ga bulk alloys and polycrystalline foams may find application in high-efficient actuators. On the other hand, the magnetization property change of a FSMA occurs under an external mechanical straining, such as compression. By measuring the magnetization or the induction voltage, the FSMAs may act as sensors, such as force, position, or acceleration sensors.

FSMAs attract much attention in the recent years as high-efficient magnetic refrigeration materials. Significant conventional or inverse magnetocaloric effects have been investigated during the martensite and magnetic transformations. For the alloys (i.e., Ni-Mn-Ga alloys) in which the martensite and magnetic transformations produce the same sign of the magnetic entropy change, the creation of the partial and full magneto-structural coupling states by compositional tuning may be adopted to optimize the magnetic entropy change and working temperature interval.

Author details

Xuexi Zhang* and Mingfang Qian

*Address all correspondence to: xxzhang@hit.edu.cn

School of Materials Science and Engineering, Harbin Institute of Technology, Harbin, China

References

[1] Chernenko VA, Vitenko IN. Structural characterization and properties of the Ni_2MnGa ribbon transforming martensitically. Materials Science Forum. 1994;**166-169**:439

[2] Kokorin VV, Chernenko VA, Valkov VI, Konoplyuk SM, Khapalyuk EA. Magnetic transformation in Ni_2MnGa compounds. Fizika Tverdogo Tela. 1995;**37**:3718

[3] Martynov VV. X-ray diffraction study of thermally and stress-induced phase transformations in single crystalline Ni-Mn-Ga alloys. Journal de Physique IV. **1995**;5:91

[4] Chernenko VA, Cesari E, Kokorin VV, Vitenko IN. The development of new ferromagnetic shape-memory alloys in Ni-Mn-Ga system. Scripta Metallurgica et Materialia. 1995;**33**:1239

[5] Ullakko K, Huang JK, Kantner C, OHandley RC, Kokorin VV. Large magnetic-field-induced strains in Ni_2MnGa single crystals. Applied Physics Letters. 1996;**69**:1966

[6] Murray SJ, Marioni M, Allen SM, O'Handley RC, Lograsso TA. 6% magnetic-field-induced strain by twin-boundary motion in ferromagnetic Ni-Mn-Ga. Applied Physics Letters. 2000;**77**:886

[7] Sozinov A, Likhachev AA, Lanska N, Ullakko K, Lindroos VK. 10% magnetic-field-induced strain in Ni-Mn-Ga seven-layered martensite. Journal de Physique IV. 2003; **112**:955

[8] Sozinov A, Lanska N, Soroka A, Zou W. 12% magnetic field-induced strain in Ni-Mn-Ga-based non-modulated martensite. Applied Physics Letters. 2013;**102**:219022

[9] Boonyongmaneerat Y, Chmielus M, Dunand DC, Müllner P. Increasing magnetoplasticity in polycrystalline Ni-Mn-Ga by reducing internal constraints through porosity. Physical Review Letters. 2007;**99**:247201

[10] Chmielus M, Zhang XX, Witherspoon C, Dunand DC, Müllner P. Giant magnetic-field-induced strains in polycrystalline Ni-Mn-Ga foams. Nature Materials. 2009;**8**:863

[11] Scheerbaum N, Heczko O, Liu J, Hinz D, Schultz L, Gutfleisch O. Magnetic field-induced twin boundary motion in polycrystalline Ni-Mn-Ga fibres. New Journal of Physics. 2008;**10**:73002

[12] Qian MF, Zhang XX, Witherspoon C, Sun JF, Müllner P. Superelasticity and shape memory effects in polycrystalline Ni-Mn-Ga microwires. Journal of Alloys and Compounds. 2013;**577**:S296

[13] Qian MF, Zhang XX, Wei LS, Martin PG, Sun JF, Geng L, Scott TB, Panina LV, Peng HX. Microstructural evolution of Ni-Mn-Ga microwires during the melt-extraction process. Journal of Alloys and Compounds. 2016;**660**:244

[14] Qian MF, Zhang XX, Wei LS, Geng L, Peng HX. Effect of chemical ordering annealing on martensitic transformation and superelasticity in polycrystalline Ni-Mn-Ga microwires. Journal of Alloys and Compounds. 2015;**645**:335

[15] Liu Y, Zhang X, Xing D, Shen H, Qian M, Liu J, Chen D, Sun J. Martensite transformation and superelasticity in polycrystalline Ni-Mn-Ga-Fe microwires prepared by melt-extraction technique. Materials Science and Engineering A. 2015;**636**:157

[16] Liu Y, Zhang X, Xing D, Qian M, Shen H, Wang H, Liu J, Sun J. Shape memory effects of $Ni_{49.7}Mn_{25.0}Ga_{19.8}Fe_{5.5}$ microwires prepared by rapid solidification. Physica Status Solidi (A). 2014;**211**:2532

[17] Zhang X, Qian M, Zhang Z, Wei L, Geng L, Sun J. Magnetostructural coupling and magnetocaloric effect in Ni-Mn-Ga-Cu microwires. Applied Physics Letters. 2016;**108**:52401

[18] Zhang H, Qian M, Zhang X, Jiang S, Wei L, Xing D, Sun J, Geng L. Magnetocaloric effect of Ni-Fe-Mn-Sn microwires prepared by melt-extraction technique. Materials & Design. 2017;**114**:1

[19] Liu Y, Zhang X, Xing D, Shen H, Chen D, Liu J, Sun J. Magnetocaloric effect (MCE) in melt-extracted Ni-Mn-Ga-Fe Heusler microwires. Journal of Alloys and Compounds. 2014;**616**:184

[20] Kainuma R, Imano Y, Ito W, Sutou Y, Morito H, Okamoto S, Kitakami O, Oikawa K, Fujita A, Kanomata T, Ishida K. Magnetic-field-induced shape recovery by reverse phase transformation. Nature. 2006;**439**:957

[21] Liu J, Gottschall T, Skokov KP, Moore JD, Gutfleisch O. Giant magnetocaloric effect driven by structural transitions. Nature Materials. 2012;**11**:620

[22] Pasquale M, Sasso CP, Lewis LH, Giudici L, Lograsso T, Schlagel D. Magnetostructural transition and magnetocaloric effect in $Ni_{55}Mn_{20}Ga_{25}$ single crystals. Physical Review B. 2005;**72**:094435

[23] Li Z, Zhang Y, Sánchez-Valdés CF, Sánchez Llamazares JL, Esling C, Zhao X, Zuo L. Giant magnetocaloric effect in melt-spun Ni-Mn-Ga ribbons with magneto-multistructural transformation. Applied Physics Letters. 2014;**104**:44101

[24] Li Z, Xu K, Zhang Y, Tao C, Zheng D, Jing C. Two successive magneto-structural transformations and their relation to enhanced magnetocaloric effect for $Ni_{55.8}Mn_{18.1}Ga_{26.1}$ Heusler alloy. Scientific Report-UK. 2015;**5**:15143

[25] Vasil'Ev AN, Bozhko AD, Khovailo VV, Dikshtein IE, Shavrov VG, Buchelnikov VD, Matsumoto M, Suzuki S, Takagi T, Tani J. Structural and magnetic phase transitions in shape-memory alloys $Ni_{2+x}Mn_{1-x}Ga$. Physical Review B. 1999;**59**:1113

[26] Chernenko VA. Compositional instability of beta-phase in Ni-Mn-Ga alloys. Scripta Materialia. 1999;**40**:523

[27] Ullakko K, Ezer Y, Sozinov A, Kimmel G, Yakovenko P, Lindroos VK. Magnetic-field-induced strains in polycrystalline Ni-Mn-Ga at room temperature. Scripta Materialia. 2001;**44**:475

[28] Murray SJ, Farinelli M, Kantner C, Huang JK, Allen SM, O'Handley RC. Field-induced strain under load in Ni-Mn-Ga magnetic shape memory materials. Journal of Applied Physics. 1998;**83**:7297

[29] Jin X, Marioni M, Bono D, Allen SM, O Handley RC, Hsu TY. Empirical mapping of Ni-Mn-Ga properties with composition and valence electron concentration. Journal of Applied Physics. 2002;**91**:8222

[30] Wu SK, Yang ST. Effect of composition on transformation temperatures of Ni-Mn-Ga shape memory alloys. Materials Letters. 2003;**57**:4291

[31] Segui C, Cesari E, Pons J, Chernenko V. Internal friction behaviour of Ni-Mn-Ga. Materials Science and Engineering A. 2004;**370**:481

[32] Chernenko VA, Pons J, Segu C, Cesari E. Premartensitic phenomena and other phase transformations in Ni-Mn-Ga alloys studied by dynamical mechanical analysis and electron diffraction. Acta Materialia. **2002**;50:53

[33] Ayuela A, Enkovaara J, Ullakko K, Nieminen RM. Structural properties of magnetic Heusler alloys. Journal of Physics: Condensed Matter. 1999;**11**:2017

[34] Jiang C, Wang J, Li P, Jia A, Xu H. Search for transformation from paramagnetic martensite to ferromagnetic austenite: NiMnGaCu alloys. Applied Physics Letters. 2009;**95**:12501

[35] Zhang XX, Witherspoon C, Mullner P, Dunand DC. Effect of pore architecture on magnetic-field-induced strain in polycrystalline Ni-Mn-Ga. Acta Materialia. 2011;**59**:2229

[36] Conde Y, Despois JF, Goodall R, Marmottant A, Salvo L, San Marchi C, Mortensen A. Replication processing of highly porous materials. Advanced Engineering Materials. 2006;**8**:795

[37] Brothers AH, Scheunemann R, DeFouw JD, Dunand DC. Processing and structure of open-celled amorphous metal foams. Scripta Materialia. 2005;**52**:335

[38] Peng H, Qin F, Phan M. Ferromagnetic Microwire Composites. DE: Springer Verlag; 2016

[39] Zhang Y, Li M, Wang YD, Lin JP, Dahmen KA, Wang ZL, Liaw PK. Superelasticity and serration behavior in small-sized NiMnGa alloys. Advanced Engineering Materials. 2014;**16**:955

[40] Zheng P, Kucza NJ, Patrick CL, Müllner P, Dunand DC. Mechanical and magnetic behavior of oligocrystalline Ni-Mn-Ga microwires. Journal of Alloys and Compounds. 2015;**624**:226

[41] Yan A, Muller KH, Gutfleisch O. Structure and magnetic entropy change of melt-spun $LaFe_{11.57}Si_{1.43}$ ribbons. Journal of Applied Physics. 2005;**97**:4494

[42] Planes A, Obrado E, GonzalezComas A, Manosa L. Premartensitic transition driven by magnetoelastic interaction in bcc ferromagnetic Ni_2MnGa. Physical Review Letters. 1997;**79**:3926

[43] Zheludev A, Shapiro SM, Wochner P, Tanner LE. Precursor effects and premartensitic transformation in Ni_2MnGa. Physical Review B. 1996;**54**:15045

[44] Brown PJ, Crangle J, Kanomata T, Matsmuoto M, Neumann KU, Ouladdiaf B, Ziebeck K. The crystal structure and phase transitions of the magnetic shape memory compound Ni_2MnGa. Journal of Physics: Condensed Matter. 2002;**14**:10159

[45] Chernenko VA, L'Vov V, Pons J, Cesari E. Superelasticity in high-temperature Ni-Mn-Ga alloys. Journal of Applied Physics. 2003;**93**:2394

[46] Mullner P, Mukherji D, Aguirre M, Erni R, Kostorz G. Micromechanics of magnetic-field-induced twin-boundary motion in Ni-Mn-Ga magnetic shape-memory alloys. Solid-Solid Phase Transformations in Inorganic Material. 2005;**2**:171

[47] Nespoli A, Besseghini S, Pittaccio S, Villa E, Viscuso S. The high potential of shape memory alloys in developing miniature mechanical devices: A review on shape memory alloy mini-actuators. Sensors and Actuators A: Physical. 2010;**158**:149

[48] Li Y, Jiang C, Liang T, Ma Y, Xu H. Martensitic transformation and magnetization of Ni-Fe-Ga ferromagnetic shape memory alloys. Scripta Materialia. 2003;**48**:1255

[49] Han ZD, Wang DH, Zhang CL, Xuan HC, Zhang JR, Gu BX, Du YW. The martensitic transformation and the magnetocaloric effect in $Ni_{50-x}Mn_{38+x}In_{12}$ alloys. Solid State Communications. 2008;**146**:124

[50] Liu Z, Wang H, Yu S, Dai X, Chen J, Wu G, Liu Y. Phase equilibrium of ferromagnetic shape memory alloy $Co_{39}Ni_{33}Al_{28}$. Scripta Materialia. 2006;**54**:1299

[51] Rolfs K, Chmielus M, Wimpory RC, Mecklenburg A, Müllner P, Schneider R. Double twinning in Ni-Mn-Ga-Co. Acta Materialia. 2010;**58**:2646

[52] Guldbakke JM, Chmielus M, Rolfs K, Schneider R, Müllner P, Raatz A. Magnetic, mechanical and fatigue properties of a $Ni_{45.4}Mn_{29.1}Ga_{21.6}Fe_{3.9}$ single crystal. Scripta Materialia. 2010;**62**:875

[53] Gao L, Dong GF, Gao ZY, Cai W. Effect of Dy addition on mechanical and magnetic properties of Mn-rich Ni-Mn-Ga ferromagnetic shape memory alloys. Journal of Alloys and Compounds. 2012;**520**:281

[54] Sozinov A, Likhachev AA, Lanska N, Ullakko K. Giant magnetic-field-induced strain in NiMnGa seven-layered martensitic phase. Applied Physics Letters. 2002;**80**:1746

[55] Cai PY, Feng SS, Chen WP, Xue SX, Li ZG, Zhou Y, Wang HB, Wang GP. Magnetic entropy change and magnetic-field-induced strain in polycrystalline $Ni_{47}Mn_{32}Ga_{21}$ alloy. Acta Physica Sinica-Chinese Edition. 2011;**60**:623

[56] Guo SH, Zhang YH, Li JL, Qi Y, Quan BY, Wang XL. Magnetic-field-induced strains of bonded Ni-Mn-Ga melt-spun ribbons. Chinese Physics Letters. 2006;**23**:227

[57] Gaitzsch U, Potschke M, Roth S, Rellinghaus B, Schultz L. Mechanical training of polycrystalline 7M $Ni_{50}Mn_{30}Ga_{20}$ magnetic shape memory alloy. Scripta Materialia. 2007;**57**:493

[58] Gaitzsch U, Potschke M, Roth S, Rellinghaus B, Schultz L. A 1% magnetostrain in polycrystalline 5M Ni-Mn-Ga. Acta Materialia. 2009;**57**:365

[59] Gaitzsch U, Romberg J, Potschke M, Roth S, Mullner P. Stable magnetic-field-induced strain above 1% in polycrystalline Ni-Mn-Ga. Scripta Materialia. 2011;**65**:679

[60] Potschke M, Weiss S, Gaitzsch U, Cong DY, Hurrich C, Roth S, Schultz L. Magnetically resettable 0.16% free strain in polycrystalline Ni-Mn-Ga plates. Scripta Materialia. 2010;**63**:383

[61] Ham-Su R, Healey JP, Underhill RS, Farrell SP, Cheng LM, Hyatt C, Chen J, Gharghouri MA. Vibration damping material employing magnetic shape memory alloys. Aerospace Materials and Manufacturing: Development, Testing, and Life Cycle Issues - Honoring William Wallace. Canadian Institute of Mining, Metallurgy and Petroleum, Montreal, QC; 2004. p. 335

[62] Wang X, Dapino MJ. Behavior of nimnga under dynamic magnetic fields considering magnetic diffusion and eddy current power loss. Proceedings of the ASME International Mechanical Engineering Congress and Exposition 2007, Vol. 10, Pts A and B: Mechanics of Solids and Structures; 2008. p 381

[63] Ganor Y, Shilo D, Shield TW, James RD. Breaching the work output limitation of ferromagnetic shape memory alloys. Applied Physics Letters. 2008;**93**:4301

[64] Scheerbaum N, Lai YW, Leisegang T, Thomas M, Liu J, Khlopkov K, McCord J, Fähler S, Träger R, Meyer DC. Constraint-dependent twin variant distribution in Ni_2MnGa single crystal, polycrystals and thin film: An EBSD study. Acta Materialia. 2010;**58**:4629

[65] Chen Y, Schuh CA. Size effects in shape memory alloy microwires. Acta Materialia. 2011;**59**:537

[66] Qian MF, Zhang XX, Wei LS, Geng L, Peng HX. Structural, Magnetic and mechanical properties of oligocrystalline Ni-Mn-Ga shape memory microwires. Materials Today: Proceedings. 2015;**2**:S577

[67] Caron L, Ou ZQ, Nguyen TT, Thanh DTC, Tegus O, Bruck E. On the determination of the magnetic entropy change in materials with first-order transitions. Journal of Magnetism and Magnetic Materials. 2009;**321**:3559

[68] Fujieda S, Fujita A, Fukamichi K. Reduction of hysteresis loss and large magnetocaloric effects in substituted compounds of itinerant-electron metamagnets $La(Fe_xSi_{1-x})_{13}$. Journal of Magnetism and Magnetic Materials. 2009;**321**:3567

[69] Zhao J, Shen J, Hu F, Li Y, Sun J, Shen B. Reduction of magnetic hysteresis loss in $La_{0.5}Pr_{0.5}Fe_{11.4}Si_{1.6}H_x$ hydrides with large magnetocaloric effects. Journal of Applied Physics. 2010;**107**:113911

[70] Shen J, Li Y, Sun J, Shen B. Effect of R substitution on magnetic properties and magnetocaloric effects of $La_{1-x}R_xFe_{11.5}Si_{1.5}$ compounds with R = Ce, Pr and Nd. Chinese Physics B. 2009;**18**:2058

[71] Pecharsky VK, Gschneidner KA. Giant magnetocaloric effect in $Gd_5(Si_2Ge_2)$. Physical Reviews Letters. 1997;**78**:4494

[72] Wada H, Tanabe Y. Giant magnetocaloric effect of $MnAs_{1-x}Sb_x$. Applied Physics Letters. 2001;**79**:3302

[73] Carvalho AMG, Coelho AA, von Ranke PJ, Alves CS. The isothermal variation of the entropy (Delta S-T) may be miscalculated from magnetization isotherms in some cases: MnAs and $Gd_5Ge_2Si_2$ compounds as examples. Journal of Alloys and Compounds. 2011;**509**:3452

[74] Sun NK, Liu F, Gao YB, Cai ZQ, Du BS, Xu SN, Si PZ. Effect of microstrain on the magnetism and magnetocaloric properties of $MnAs_{0.97}P_{0.03}$. Applied Physics Letters. 2012; **100**:112407

[75] Nguyen HD, Zhang L, Ou ZQ, Bruck E. Magnetoelastic coupling and magnetocaloric effect in hexagonal Mn-Fe-P-Si compounds. Scripta Materialia. 2012;**67**:975

[76] Yibole H, Guillou F, Zhang L, van Dijk NH, Bruck E. Direct measurement of the magnetocaloric effect in MnFe(P,X) (X = As, Ge, Si) materials. Journal of Physics D: Applied Physics. 2014;**47**:75002

[77] Manekar M, Roy SB. Reproducible room temperature giant magnetocaloric effect in Fe-Rh. Journal of Physics D: Applied Physics. 2008;**41**:192004

[78] Krenke T, Duman E, Acet M, Wassermann EF, Moya X, Manosa L, Planes A. Inverse magnetocaloric effect in ferromagnetic Ni-Mn-Sn alloys. Nature Materials. **2005**;4:450

[79] Krenke T, Duman E, Acet M, Moya X, Manosa L, Planes A. Effect of Co and Fe on the inverse magnetocaloric properties of Ni-Mn-Sn. Journal of Applied Physics. 2007; **102**:33903

[80] Gao B, Hu FX, Shen J, Wang J, Sun JR, Shen BG. Field-induced structural transition and the related magnetic entropy change in $Ni_{43}Mn_{43}Co_3Sn_{11}$ alloy. Journal of Magnetism and Magnetic Materials. 2009;**321**:2571

[81] Takeuchi AY, Guimaraes CE, Passamani EC, Larica C. Enhancement of magnetocaloric properties near room temperature in Ga-doped $Ni_{50}Mn_{34.5}In_{15.5}$ Heusler-type alloy. Journal of Applied Physics. 2012;**111**:3867

[82] Zhang Y, Zheng Q, Xia W, Zhang J, Du J, Yan A. Enhanced large magnetic entropy change and adiabatic temperature change of $Ni_{43}Mn_{46}Sn_{11}$ alloys by a rapid solidification method. Scripta Materialia. 2015;**104**:41

[83] Yan A, Mueller KH, Gutfleisch O. Magnetocaloric effect in $LaFe_{11.8-x}Co_xSi_{1.2}$ melt-spun ribbons. Journal of Alloys and Compounds. 2008;**450**:18

[84] Zhang X, Qian M, Miao S, Su R, Liu Y, Geng L, Sun J. Enhanced magnetic entropy change and working temperature interval in Ni-Mn-In-Co alloys. Journal of Alloys and Compounds. 2016;**656**:154

[85] Zhang Y, Zhang L, Zheng Q, Zheng X, Li M, Du J, Yan A. Enhanced magnetic refrigeration properties in Mn-rich Ni-Mn-Sn ribbons by optimal annealing. Scientific Reports-UK. 2015;**5**

[86] Dubenko I, Samanta T, Pathak AK, Kazakov A, Prudnikov V, Stadler S, Granovsky A, Zhukov A, Ali N. Magnetocaloric effect and multifunctional properties of Ni-Mn-based Heusler alloys. Journal of Magnetism and Magnetic Materials. 2012;**324**:3530

[87] Quetz A, Koshkid'Ko YS, Titov I, Rodionov I, Pandey S, Aryal A, Ibarra-Gaytan PJ, Prudnikov V, Granovsky A, Dubenko I, Samanta T, Cwik J, Sanchez Llamazares JL, Stadler S, Lahderanta E, Ali N. Giant reversible inverse magnetocaloric effects in $Ni_{50}Mn_{35}In_{15}$ Heusler alloys. Journal of Alloys and Compounds. 2016;**683**:139

[88] Dong JD, Yan AR, Liu J. Microstructure and magnetocaloric properties of melt-extracted La-Fe-Si microwires. Journal of Magnetism and Magnetic Materials. 2014;**357**:73

[89] Ilyn MI, Zhukova V, Santos JD, Sánchez ML, Prida VM, Hernando B, Larin V, González J, Tishin AM, Zhukov A. Magnetocaloric effect in nanogranular glass coated microwires. Physica Status Solidi (A). 2008;**205**:1378

[90] Aguilar-Ortiz CO, Soto-Parra D, Álvarez-Alonso P, Lázpita P, Salazar D, Castillo-Villa PO, Flores-Zúñiga H, Chernenko VA. Influence of Fe doping and magnetic field on martensitic transition in Ni-Mn-Sn melt-spun ribbons. Acta Materialia. 2016;**107**:9

[91] Yuzuak E, Dincer I, Elerman Y, Auge A, Teichert N, Hutten A. Inverse magnetocaloric effect of epitaxial Ni-Mn-Sn thin films. Applied Physics Letters. 2013;**103**:222403

[92] Recarte V, Perez-Landazabal JI, Sanchez-Alarcos V, Chernenko VA, Ohtsuka M. Magnetocaloric effect linked to the martensitic transformation in sputter-deposited Ni-Mn-Ga thin films. Applied Physics Letters. 2009;**95**

[93] Hu F, Shen B, Sun J. Magnetic entropy change involving martensitic transition in NiMn-based Heusler alloys. Chinese Physics B. 2013;**22**:037505

[94] Duan J, Huang P, Zhang H, Long Y, Wu G, Ye R, Chang Y, Wan F. Negative and positive magnetocaloric effect in Ni-Fe-Mn-Ga alloy. Journal of Magnetism and Magnetic Materials. 2007;**309**:96

[95] Varga R, Ryba T, Vargova Z, Saksl K, Zhukova V, Zhukov A. Magnetic and structural properties of Ni-Mn-Ga Heusler-type microwires. Scripta Materialia. 2011;**65**:703

[96] Srivastava SK, Srivastava VK, Chatterjee R. Magnetocaloric properties of $Ni_{2.29}Mn_{0.89}Ga_{0.82}$ alloy. Solid State Communications. 2012;**152**:372

[97] Stadler S, Khan M, Mitchell J, Ali N, Gomes AM, Dubenko I, Takeuchi AY, Guimarães AP. Magnetocaloric properties of $Ni_2Mn_{1-x}Cu_xGa$. Applied Physics Letters. 2006;**88**:192511

[98] Huang C, Wang Y, Tang Z, Liao X, Yang S, Song X. Influence of atomic ordering on elastocaloric and magnetocaloric effects of a Ni-Cu-Mn-Ga ferromagnetic shape memory alloy. Journal of Alloys and Compounds. 2015;**630**:244

[99] Duan JF, Long Y, Bao B, Zhang H, Ye RC, Chang YQ, Wan FR, Wu GH. Experimental and theoretical investigations of the magnetocaloric effect of $Ni_{2.15}Mn_{0.85-x}Cu_xGa$ ($x = 0.05, 0.07$) alloys. Journal of Applied Physics. 2008;**103**

[100] Dan'Kov SY, Tishin AM, Pecharsky VK, Gschneidner KA. Magnetic phase transitions and the magnetothermal properties of gadolinium. Physical Review B. 1998;**57**:3478

[101] Gutfleisch O, Yan A, Müller KH. Large magnetocaloric effect in melt-spun $LaFe_{13-x}Si_x$. Journal of Applied Physics. 2005;**97**:10M

[102] Xu HB, Ma YQ, Jiang CB. A high-temperature shape-memory alloy $Ni_{54}Mn_{25}Ga_{21}$. Applied Physics Letters. 2003;**82**:3206

[103] Xin Y, Li Y, Chai L, Xu HB. Shape memory characteristics of dual-phase Ni-Mn-Ga based high temperature shape memory alloys. Scripta Materialia. 2007;**57**:599

[104] Glavatskyy I, Glavatska N, Soderberg O, Hannula SP, Hoffmann JU. Transformation temperatures and magnetoplasticity of Ni-Mn-Ga alloyed with Si, In, Co or Fe. Scripta Materialia. 2006;54:1891

[105] Glavatskyy I, Glavatska N, Dobrinsky A, Hoffmann JU, Sorberg O, Hannula SP. Crystal structure and high-temperature magnetoplasticity in the new Ni-Mn-Ga-Cu magnetic shape memory alloys. Scripta Materialia. 2007;**56**:565

[106] Ma YQ, Yang SY, Wang CP, Liu XJ. Tensile characteristics and shape memory effect of $Ni_{56}Mn_{21}Co_4Ga_{19}$ high-temperature shape memory alloy. Scripta Materialia. 2008;**58**:918

[107] Soto D, Hernandez FA, Flores-Zuniga H, Moya X, Manosa L, Planes A, Aksoy S, Acet M, Krenke T. Phase diagram of Fe-doped Ni-Mn-Ga ferromagnetic shape-memory alloys. Physical Review B. 2008;**77**:1

[108] Soto-Parra DE, Vives E, Gonzalez-Alonso D, Manosa L, Planes A, Romero R, Matutes-Aquino JA, Ochoa-Gamboa RA, Flores-Zuniga H. Stress- and magnetic field-induced entropy changes in Fe-doped Ni-Mn-Ga shape-memory alloys. Applied Physics Letters. 2010;**96**

[109] Magnetic shape memory actuators. http://www.themsmnet.net/devices.html

[110] Wilson SA, Jourdain RPJ, Zhang Q, Dorey RA, Bowen CR, Willander M, Wahab QU, Willander M, Al-hilli SM, Nur O, Quandt E, Johansson C, Pagounis E, Kohl M, Matovic J, Samel B, van der Wijngaart W, Jager EWH, Carlsson D, Djinovic Z, Wegener M, Moldovan C, Iosub R, Abad E, Wendlandt M, Rusu C, Persson K. New materials for micro-scale sensors and actuators. Materials Science and Engineering: R: Reports. 2007;**56**:1

[111] Boise State Micro-Pump Aids Neurological Research. 2014. https://news.boisestate.edu/update/2014/09/24/micro-pump-developed-boise-state-aids-neurological-research-england/

[112] Franco V, Blazquez JS, Ingale B, Conde A. The magnetocaloric effect and magnetic refrigeration near room temperature: Materials and models. In: Clarke DR, editor. Annual Review of Materials Research, vol. 42. Palo Alto: Annual Reviews; 2012. p. 305

[113] Krautz M, Skokov K, Gottschall T, Teixeira CS, Waske A, Liu J, Schultz L, Gutfleisch O. Systematic investigation of Mn substituted La(Fe,Si)(13) alloys and their hydrides for room-temperature magnetocaloric application. Journal of Alloys and Compounds. 2014;**598**:27

Chapter 2

Shape Memory Wires in R^3

Shinya Okabe, Takashi Suzuki and Shuji Yoshikawa

Additional information is available at the end of the chapter

http://dx.doi.org/10.5772/intechopen.69175

Abstract

We propose a new model describing the dynamics of wire made of shape memory alloys, by combining an elastic curve theory and the Ginzburg-Landau theory. The wire is assumed to be a closed curve and is not to be stretched with deformation. The derived system of nonlinear partial differential equations consists of a thermoelastic system and a geometric evolution equation under the inextensible condition. We also show that the system has dual variation structure as well as a straight material case. The structure implies stability of infinitesimally stable stationary state in the Lyapunov sense.

Keywords: shape memory alloys, elastic curve, thermoelastic system, nonlinear partial differential equations, Ginzburg-Landau theory, phase transition, stability, dual variation principle

1. Introduction

Shape memory effect arises from the phase transition of lattice structure. Although there are many models for shape memory alloys, one of the classical model is proposed by Falk, which we call the **Falk model**. Falk applied the Ginzburg-Landau theory for phase transition to shape memory alloys by regarding shear strain ϵ as an order parameter (see e.g., [1]). That is, the Helmholz free energy density proposed by Falk is given by

$$\begin{aligned}\tilde{f}(\epsilon,\theta,\partial_x^2 u) &:= \frac{1}{2}|\partial_x^2 u|^2 + f(\epsilon,\theta) + f_0(\theta)\\ &= \frac{1}{2}|\partial_x^2 u|^2 + \left(\frac{1}{6}|\epsilon|^6 - \frac{1}{4}|\epsilon|^4 + \frac{1}{2}(\theta-\theta_c)|\epsilon|^2\right) + f_0(\theta)\end{aligned} \tag{1}$$

where u and θ are displacement and absolute temperature, respectively, and a positive constant θ_c denotes the critical temperature of the phase transition. We call the first term $\frac{|\partial_x^2 u|^2}{2}$ **curvature energy density**, the second term $f(\epsilon, \theta)$ **nonlinear elastic energy density** and the third term $f_0(\theta)$ **thermal energy density**. In other words, Falk represented the phase transition by using the form of the nonlinear elastic energy density. We also remark that for simplicity, all physical constants without the critical temperature are normalized by unity. The Falk model was proposed for straight materials. In the model, the material is built up by a stack of layers parallel to the so-called **habit plane** (see [2]) and assumed that the displacement u in that direction to depend only on a coordinate x perpendicular to the habit plane, that is, the variable x runs in the stacking direction. Then, the material conserves its volume.

From the point of the small deformation theory, we may use a linearized approximation relation $\epsilon = \partial_x u$. Moreover, we take f_0 as the following typical form:

$$f_0(\theta) := \theta - \theta \log\theta.$$

Then following a standard procedure of derivation of thermoelastic system (see e.g., [3]), we can derive the system of nonlinear partial differential equations called the **Falk model**:

$$\begin{cases}\partial_t^2 u + \partial_x^4 u = \partial_x\{(\partial_x u)^5 - (\partial_x u)^3 + (\theta-\theta_c)\partial_x u\},\\ \partial_t\theta - \partial_x^2\theta = \theta\partial_x u\,\partial_t\partial_x u.\end{cases} \tag{2}$$

Here, unknowns are displacement u and absolute temperature θ, and ∂_x and ∂_t represent partial differential operator with respect to x and t, respectively. The model is well known as one of classical models describing shape memory alloys. For the other models, we refer Fremond [4], Fremond-Miyazaki [5] and reference therein. The Falk model (Eq. (2)) has been studied actively in the mathematical literature. In the isothermal case, well-posedness, stability of solitary-wave solution, existence of steady state, travelling wave solution and invariant measure have been investigated by Fang-Grillakis [6], Falk-Laedke-Spatschek [7], Friedman-Sprekels [8], Garcke [9] and Tsutsumi-Yoshikawa [10], respectively. For the full system (Eq. (1)), the well-posedness results are found in, for example, [11–13] and so on, and numerical results are found in Hoffmann-Zou [14], Niezgodka-Sprekels [15] by finite element method and in Matus-Melnik-Wang-Rybak [16] and Yoshikawa [17, 18] by the finite difference method. In particular, in Ref. [19], the stability of steady state in the Lyapunov sense was shown. More precisely, the stationary state of Eq. (2) is expressed as a nonlocal nonlinear elliptic problem. If there exists a linearized stable critical point for the functional corresponding to the elliptic problem, then for each neighbourhood U of the equilibrium, we can find a neighbourhood W of the equilibrium such that the solution of Eq. (2) with the initial data in W stays in U for any time. The proof can be shown by the **dual variation principle** which appears in most of the models in non-equilibrium statistical thermodynamics (see [20]).

http://dx.doi.org/10.5772/intechopen.69175

The existence of several non-trivial steady states for low-temperature phase and low-energy case is proved in Ref. [21]. The numerical simulation given in Ref. [17] exactly indicates the properties mentioned above. The dual variation structure appears also in a multi-dimensional case [22]; however, well-posedness of the multi-dimensional model corresponding to Eq. (2) is still open in large initial data case due to the propagation of singularity. That is one of our motivations of this problem.

We mention mathematical studies on the motion of curves governed by geometric evolution equations. One of the typical objects is curve-shortening flow derived as an L^2 gradient flow for the length functional of curve γ:

$$L(\gamma) := \int_\gamma ds$$

where s denotes the arc length parameter of γ. By Gage [23], Gage and Hamilton [24] and Grayson [25], it is well known that the curve-shortening flow shrinks simple closed curves to a point in a finite time. Since the curve-shortening flow can be regarded as a one-dimensional case of mean curvature flow for surfaces, the flow is applicable to various mathematical analysis. For example, the curve-shortening flow plays an important role in studies on phase transition. We also mention the curve-straightening flow which has been attracted a great interest and studied actively in mathematical literature. The flow is derived as an L^2 gradient flow for the elastic energy

$$K(\gamma) := \frac{1}{2}\int_\gamma \kappa^2 \, ds$$

where κ denotes the scalar curvature of γ. It is well known that the flow is applicable to studies on elastic curve inspired by Bernoulli and Euler. Indeed, the curve-straightening flow under the length constraint $L(\gamma) \equiv C$ converges to a classical elastic curve so-called elastica. There is also an interest in the study on motion of curves governed by the L^2-gradient flow for E under the inextensible condition. Under the condition, the length constraint $L(\gamma) \equiv C$ is also satisfied for the condition means that the curve does not stretch. As we will state in Section 2.2, the constraint is imposed on each point of curves. Thus, a standard Lagrange multiplier theory does not work. Therefore, we have to make use of geometric properties of curves governed by the flow.

The purpose of this chapter is to derive a mathematical model describing thermoelastic deformation of shape memory wire in $\boldsymbol{R}^3$. In particular, we regard the wire as a closed space curve satisfying the inextensible condition. From the physical point of view, it may be unnatural that the wire does not stretch. However, the contribution of this chapter is to adopt a geometric analysis into a classical thermoelastic theory with phase transition inspired by Falk.

2. Setting and derivation of equations

We denote the closed curve representing shape of wire by $\Gamma = \{\gamma(\xi) : \xi \in \Xi\}$, where the variable ξ is an arbitrary parameter not necessarily the arc length parameter. Let us define a displacement vector from a point ξ in an original shape $\Gamma^0 = \{\gamma^0(\xi) : \xi \in \Xi\}$ by $u(\xi)$ (see **Figure 1**); namely, it holds that

$$\gamma(\xi) = \gamma^0(\xi) + u(\xi)$$

for

- $\gamma(\xi) := \left(\gamma_1(\xi),\ \gamma_2(\xi),\ \gamma_3(\xi)\right)$: vector representing the shape,
- $\gamma^0(\xi) := \left(\gamma_1^0(\xi), \gamma_2^0(\xi), \gamma_3^0(\xi)\right)$: vector representing the original shape,
- $u(\xi) := \left(u_1(\xi), u_2(\xi), u_3(\xi)\right)$: displacement vector.

Throughout this chapter, we denote by L the length of Γ^0, and hence, the length of Γ is also L from the non-stretching assumption. To apply the idea by Falk, we need to determine the form of strain and free energy (Eq. (1)) suitable for this setting.

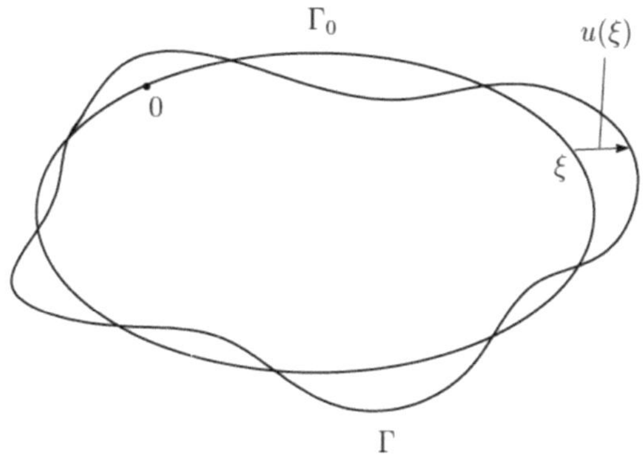

Figure 1. Original shape Γ_0 and deformed curve Γ.

2.1. Definition of strain

We first consider the strain. Let $\gamma^0(\xi)$ be a space closed curve, where ξ is a parameter (not necessary to be the arc length parameter). For γ^0 (ξ), we define the displacement vector by $u(\xi)$, and we denote $\gamma(\xi) = \gamma^0(\xi) + u(\xi)$. Since the relation "strain ≈ line element" holds, let us first pursue line element between γ^0 (ξ) and γ (ξ). From the direct calculation, we have

$$\begin{aligned} |\gamma'(\xi)|^2 - |\gamma^{0'}(\xi)|^2 &= \{\gamma^{0'}(\xi) + u'(\xi)\} \cdot \{\gamma^{0'}(\xi) + u'(\xi)\} - \gamma^{0\prime}(\xi) \cdot \gamma^{0'}(\xi) \\ &= 2\gamma^{0'}(\xi) \cdot u'(\xi) + |u'(\xi)|^2. \end{aligned}$$

Here, if we assume a smallness of deformation, then we may assume

$$|\gamma^{0'}(\xi) \cdot u'(\xi)| \gg |u'(\xi)|^2.$$

From now on, we regard the strain as

$$\gamma^{0'}(\xi) \cdot u'(\xi).$$

2.2. Definition of energy functional

Let $\gamma^0(\xi)$ be an initial closed curve and $\gamma(\xi, t)$ denote a family of closed curves starting from $\gamma^0(\xi)$. Recalling that the arc length parameter $s(\xi, t)$of $\gamma(\xi, t)$ is given by

$$ds = \left|\frac{\partial \gamma}{\partial \xi}\right| d\xi,$$

we can write kinetic energy $M(\gamma)$ of γ as

$$M(\gamma) := \oint \left|\frac{\partial \gamma}{\partial t}\right|^2 \left|\frac{\partial \gamma}{\partial \xi}\right| d\xi.$$

In a similar manner, thermal energy is defined by

$$F_0(\gamma) := \oint f_0(\xi) \left|\frac{\partial \gamma}{\partial \xi}\right| d\xi,$$

and the curvature energy is expressed as

$$K(\gamma) := \oint \kappa^2 \left|\frac{\partial \gamma}{\partial \xi}\right| d\xi.$$

Observe that the scalar curvature κ is written as

$$\kappa = \left\{ \left|\frac{\partial \gamma}{\partial \xi}\right| \frac{\partial^2 \gamma}{\partial \xi^2} - \left(\frac{\partial \gamma}{\partial \xi} \cdot \frac{\partial^2 \gamma}{\partial \xi^2}\right) \frac{\partial \gamma}{\partial \xi} \right\} \left|\frac{\partial \gamma}{\partial \xi}\right|^{-3}.$$

Lastly, the nonlinear elastic energy density is given by

$$f(\partial_\xi \gamma,\ \theta; \partial_\xi \gamma^0) := \frac{1}{6}\left(\frac{\partial \gamma^0}{\partial \xi} \cdot \frac{\partial u}{\partial \xi}\right)^6 - \frac{1}{4}\left(\frac{\partial \gamma^0}{\partial \xi} \cdot \frac{\partial u}{\partial \xi}\right)^4 + \frac{1}{2}(\theta - \theta_c)\left(\frac{\partial \gamma^0}{\partial \xi} \cdot \frac{\partial u}{\partial \xi}\right)^2$$

and then the nonlinear elastic energy is written as

$$F(\partial_\xi \gamma, \theta; \partial_\xi \gamma^0) := \oint f(\partial_\xi \gamma,\ \theta; \partial_\xi \gamma^0) \left|\frac{\partial \gamma}{\partial \xi}\right| d\xi$$

where $u = \gamma - \gamma^0$. Thus, we obtain the Helmholtz energy for our setting as

$$H(\gamma, \theta, \gamma^0) := M(\gamma) + K(\gamma) + F(\partial_\xi \gamma, \theta; \partial_\xi \gamma^0) + F_0(\theta).$$

From now on, let $s \in \boldsymbol{R}/L\boldsymbol{Z} =: S_L^1$ be the arc length parameter of the initial closed curve $\gamma^0 = \gamma^0(s)$. It follows from the property of arc length parameter that $|\gamma^{0'}(s)| \equiv 1$. In a similar fashion to the above equation, $\gamma(s, t)$ means the closed curve deformed along evolution from $\gamma^0(s)$. Moreover, in what follows, we assume that $\gamma(s, t)$ satisfies

$$|\partial_s \gamma(s,t)| \equiv 1 \tag{3}$$

which means "s is arc length parameter of γ not only the initial time but also every time t". From the assumption, we can rewrite M, F_0 and F shortly as

$$M(\gamma) = \int_0^L | \partial_t \gamma |^2 \, ds,$$

$$F_0(\theta) = \int_0^L f_0(\theta) ds,$$

$$F(\partial_s \gamma, \theta; \partial_s \gamma^0) = \int_0^L f(\partial_s \gamma, \theta; \partial_s \gamma^0) \, ds.$$

Moreover, since $\partial_s \gamma \cdot \partial_s^2 \gamma = 0$, from Eq. (3), it holds that

$$K(\gamma) = \int_0^L | \partial_s^2 \gamma |^2 ds.$$

Therefore, the Helmholtz free energy density is denoted by

$$H(t) := \frac{1}{2} ||\partial_t \gamma(\cdot, t)||^2_{L^2(S^1_L)} + \frac{1}{2} ||\partial_s^2 \gamma(\cdot, t)||^2_{L^2(S^1_L)} + F\left(\partial_s \gamma(\cdot, t),\ \theta(\cdot, t); \partial_s \gamma^0(\cdot)\right) + F_0\left(\theta(\cdot, t)\right).$$

We will explain that for the free energy under some assumptions, the following system of nonlinear partial differential equations is derived:

$$\begin{cases} \partial_t^2 \gamma + \partial_s^4 \gamma + \partial_s f_{,\partial_s \gamma}(\partial_s \gamma,\ \theta; \partial_s \gamma^0) - \partial_s \left\{ (v - 2\, |\partial_s^2 \gamma|^2)\, \partial_s \gamma \right\} = 0, \\ -\partial_s^2 v + |\partial_s^2 \gamma|^2 v = 2\, |\partial_s^2 \gamma|^4 - |\partial_s^3 \gamma|^2 + |\partial_s \partial_t \gamma|^2 + \partial_s^2 f_{,\partial_s \gamma}(\partial_s \gamma,\ \theta; \partial_s \gamma^0) \cdot \partial_s \gamma, \\ \partial_t \theta - \partial_s^2 \theta = \theta\left(\partial_s \gamma^0 \cdot \partial_s (\gamma - \gamma^0)\right)(\partial_t \partial_s \gamma \cdot \partial_s \gamma^0) \end{cases} \tag{4}$$

where

$$f_{,\partial_s \gamma}(\partial_s \gamma,\ \theta; \partial_s \gamma^0) = \left(\frac{\partial f}{\partial_s \gamma_1}, \frac{\partial f}{\partial_s \gamma_2}, \frac{\partial f}{\partial_s \gamma_3} \right)$$
$$= \{ (\partial_s \gamma^0 \cdot \partial_s u)^5 - (\partial_s \gamma^0 \cdot \partial_s u)^3 + (\theta - \theta_c) \partial_s \gamma^0 \cdot \partial_s u \} \partial_s \gamma^0.$$

2.3. Equation of motion

By using the Hamilton principle, we derive an equation of the motion of γ. Namely, we will derive the Euler-Lagrange equation for the functional:

$$\tilde{H}(\gamma,\ \theta; \gamma^0) = \int_{t_2}^{t_1} \left\{ \frac{1}{2} ||\partial_t \gamma(\cdot, t)||^2_{L^2(S^1_L)} + \frac{1}{2} ||\partial_s^2 \gamma(\cdot, t)||^2_{L^2(S^1_L)} - F\left(\partial_s \gamma(\cdot, t),\ \theta(\cdot, t); \partial_s \gamma^0(\cdot)\right) - F_0\left(\theta(\cdot, t)\right) \right\} dt. \tag{5}$$

Let us denote the variation of γ by

$$\gamma(s, t; \theta) := \gamma(s, t) + \varepsilon\, \varphi(s, t)$$

where ε is a sufficiently small positive parameter and φ is sufficiently smooth and satisfies $\varphi(s, t_1) = \varphi(s, t_2) = 0$. Moreover, from the assumption Eq. (3), it is also necessary to hold that

$$\frac{d}{d\varepsilon}|\partial_s\gamma(s, t; \varepsilon)|\Big|_{\varepsilon=0} = 0.$$

Since

$$\frac{d}{d\varepsilon}|\partial_s\gamma(s, t; \theta)|\Big|_{\varepsilon=0} = \partial_s\gamma(s, t)\cdot\partial_s\varphi(s, t)$$

φ has to satisfy

$$\partial_s\gamma(s, t)\cdot\partial_s\varphi(s, t) = 0$$

for any $s \in S_L^1$ and $t > 0$. Calculating the first variation of the energy functional, we have

$$\frac{d}{d\varepsilon}\tilde{H}(\gamma,\ \theta; \gamma^0)\Big|_{\varepsilon=0} = \int_{t_1}^{t_2}\{\langle\partial_t\gamma, \partial_t\varphi\rangle - \langle\partial_s^2\gamma, \partial_s^2\varphi\rangle - \langle f_{,\partial_s\gamma}(\partial_s\gamma, \theta; \partial_s\gamma^0), \partial_s\varphi\rangle\}dt.$$

From the integral by parts, the right-hand side is rewritten as follows:

$$-\int_{t_1}^{t_2}\langle\partial_t^2\gamma + \partial_s^4\gamma - \partial_s f_{,\partial_s\gamma}(\partial_s\gamma, \theta; \partial_s\gamma^0), \varphi\rangle dt. \tag{6}$$

Then the integral Eq. (6) is equal to 0 for any φ satisfying $\varphi(s, t_1) = \varphi(s, t_2) = 0$ and $\partial_s\gamma\cdot\partial_s\varphi \equiv 0$. For the purpose, we define

$$V := \{\varphi \mid \partial_s\gamma\cdot\partial_s\varphi \equiv 0\}.$$

The orthogonal complement $V^\perp$ of the space V with respect to $L^2(S_L^1)$ inner product is given by

$$V^\perp = \{\partial_s(w\partial_s\gamma)w = w(s, t) \text{ is a scalar function}\}. \tag{7}$$

Here, we remark that in the case where γ is a curve embedded in three-dimensional space (not a planar curve), $\partial_s^2\gamma \neq \mathbf{0}$ has to be satisfied for every $(s, t) \in S_L^1 \times \mathbf{R}_+$. In the end of this section, we will show the reason why $V^\perp$ is given as above. Consequently, if for the direction $\partial_t^2\gamma + \partial_s^4\gamma - \partial_s f_{,\partial_s\gamma}(\partial_s\gamma, \theta; \partial_s\gamma^0)$, there exists a scalar function $w = w(s, t)$ such that

$$\partial_t^2\gamma + \partial_s^4\gamma - \partial_s f_{,\partial_s\gamma}(\partial_s\gamma, \theta; \partial_s\gamma^0) = \partial_s(w\partial_s\gamma) \tag{8}$$

then Eq. (6) is equal to 0.

Next, we derive the equation for w. From the assumption Eq. (3), we see that

$$0 = \partial_t^2|\partial_s\gamma|^2 = 2\partial_s\partial_t^2\gamma \cdot \partial_s\gamma + 2|\partial_s\partial_t\gamma|^2$$

It follows from Eq. (8) that

$$\{-\partial_s^5\gamma + \partial_s^2 f_{,\partial_s\gamma}(\partial_s\gamma, \theta; \partial_s\gamma^0) + \partial_s^2(w\partial_s\gamma)\} \cdot \partial_s\gamma = -|\partial_s\partial_t\gamma|^2. \tag{9}$$

Differentiating Eq. (3), we obtain

$$\begin{aligned}
\partial_s\gamma \cdot \partial_s^2\gamma &= 0, \\
\partial_s\gamma \cdot \partial_s^3\gamma &= -|\partial_s^2\gamma|^2, \\
\partial_s\gamma \cdot \partial_s^4\gamma &= -\frac{3}{2}\partial_s(|\partial_s^2\gamma|^2), \\
\partial_s\gamma \cdot \partial_s^5\gamma &= -\ 2\partial_s^2(|\partial_s^2\gamma|^2) +\ |\partial_s^3\gamma|^2.
\end{aligned}$$

By the relations, we will rewrite Eq. (9). It follows from the direct calculation that

$$\partial_s^2\ (w\partial_s\gamma) \cdot \partial_s\gamma = \partial_s^2 w\ -\ w|\partial_s^2\gamma|^2.$$

Since from the definition

$$\begin{aligned}
&f_{,\partial_s\gamma}(\partial_s\gamma,\ \theta; \partial_s\gamma^0) \\
&= \left\{\left(\partial_s\gamma^0 \cdot \partial_s(\gamma - \gamma^0)\right)^5 - \left(\partial_s\gamma^0 \cdot \partial_s(\gamma - \gamma^0)\right)^3 + (\theta - \theta_c)\left(\partial_s\gamma^0 \cdot \partial_s(\gamma - \gamma^0)\right)\right\}\partial_s\gamma^0
\end{aligned}$$

we also obtain

$$\begin{aligned}
&\partial_s^2 f_{,\partial_s\gamma}(\partial_s\gamma,\ \theta; \partial_s\gamma^0) \cdot \partial_s\gamma \\
&= \partial_s^2\left[\left\{\left(\partial_s\gamma^0 \cdot \partial_s(\gamma - \gamma^0)\right)^5 - \left(\partial_s\gamma^0 \cdot \partial_s(\gamma - \gamma^0)\right)^3 + (\theta - \theta_c)\left(\partial_s\gamma^0 \cdot \partial_s(\gamma - \gamma^0)\right)\right\}\partial_s\gamma^0\right] \cdot \partial_s\gamma.
\end{aligned}$$

Therefore, substituting these into Eq. (9), we find

$$\begin{aligned}
0 = 2\partial_s^2(|\partial_s^2\gamma|^2) - |\partial_s^3\gamma|^2 + \partial_s^2 w - w|\partial_s^2\gamma|^2 + |\partial_s\partial_t\gamma|^2 + \partial_s^2\Big[\Big\{\left(\partial_s\gamma^0 \cdot \partial_s(\gamma - \gamma^0)\right)^5 \\
-\left(\partial_s\gamma^0 \cdot \partial_s\ (\gamma - \gamma^0)\right)^3 + (\theta - \theta_c)\left(\partial_s\ \gamma^0 \cdot \partial_s\ (\gamma - \gamma^0)\right)\Big\}\partial_s\gamma^0\Big] \cdot \partial_s\gamma.
\end{aligned}$$

Here setting the new unknown v by $v := w + 2|\partial_s^2\gamma|^2$, we can rewrite the equation as follows:

$$-\partial_s^2 v + |\partial_s^2\gamma|^2 v = 2\,|\partial_s^2\gamma|^4 - |\partial_s^3\gamma|^2 + |\partial_s\partial_t\gamma|^2$$
$$+\partial_s^2\left[\left\{\left(\partial_s\gamma^0\cdot\partial_s(\gamma-\gamma^0)\right)^5 - \left(\partial_s\gamma^0\cdot\partial_s(\gamma-\gamma^0)\right)^3 + (\theta-\theta_c)\left(\partial_s\gamma^0\cdot\partial_s(\gamma-\gamma^0)\right)\right\}\partial_s\gamma^0\right]\cdot\partial_s\gamma.$$

Consequently, under given temperature θ, the equation of motion is given by

$$\begin{cases} \partial_t^2\gamma + \partial_s^4\gamma - \partial_s f_{,\partial_s\gamma}(\partial_s\gamma,\theta;\partial_s\gamma^0) - \partial_s\{(v-2|\partial_s^2\gamma|^2)\partial_s\gamma\} = 0, \\ -\partial_s^2 v + |\partial_s^2\gamma|^2 v = 2|\partial_s^2\gamma|^4 - |\partial_s^3\gamma|^2 + |\partial_s\partial_t\gamma|^2 - \partial_s^2 f_{,\partial_s\gamma}(\partial_s\gamma,\ \theta;\partial_s\gamma^0)\cdot\partial_s\gamma. \end{cases}$$

At the rest of this section, we prove that the orthogonal complement of V is given by Eq. (7).

Lemma 1. *Let $\gamma(s,t)$ be a smooth curve in $\mathbf{R}^3$ and $s\in S_L^1$ be an arc length parameter of γ for any t. Suppose that $\partial_s^2\gamma \neq \mathbf{0}$ holds for all $(s,t)\in S_L^1\times\mathbf{R}_+$ then the orthogonal complement $V^\perp$ of the space $V=\{\varphi|\partial_s\gamma\cdot\partial_s\varphi\equiv 0\}$ with respect to $L^2(S_L^1)$ inner product is represented by*

$$V^\perp = \{\partial_s(w\partial_s\gamma) w = w(s,t) \text{ is a scalar function}\}.$$

Proof. Observe that $\partial_s\gamma$, $\partial_s^2\gamma$ and the outer product $\partial_s\gamma\times\partial_s^2\gamma$ are orthogonal each other. Under the assumption $\partial_s^2\gamma\neq\mathbf{0}$, a coordinate system defined on γ consists of the vectors. Then arbitrary vector $\eta = \eta(s,t)$ can be represented as

$$\eta(s,t) = \eta_1(s,t)\partial_s\gamma(s,t) + \eta_2(s,t)\,\partial_s^2\gamma(s,t) + \eta_3(s,t)\,\partial_s\gamma(s,t)\times\partial_s^2\gamma(s,t).$$

If we assume additionally $\eta\in V$, then we obtain

$$\begin{aligned} 0 &= \partial_s\eta\cdot\partial_s\gamma \\ &= \{\partial_s\eta_1\partial_s\gamma + \eta_1\partial_s^2\gamma + \partial_s\eta_2\partial_s^2\gamma + \eta_2\partial_s^3\gamma + \partial_s\eta_3\partial_s\gamma\times\partial_s^2\gamma + \eta_3\,(\partial_s^2\gamma\times\partial_s^2\gamma + \partial_s\gamma\times\partial_s^3\gamma)\}\cdot\partial_s\gamma \\ &= \partial_s\eta_1 + \eta_2\partial_s\gamma\cdot\partial_s^3\gamma \\ &= \partial_s\eta_1 - \eta_2|\partial_s^2\gamma|^2 \end{aligned}$$

that is

$$\partial_s\eta_1 = |\partial_s^2\gamma|^2\eta_2. \tag{10}$$

In other words, an element of V consists of η_1, η_2 satisfying Eq. (10) and arbitrary η_3. However, we remark that we cannot take η_2 freely. Indeed, in order to verify L periodicity of η_1, η_2, we have to show the following condition:

$$\int_0^L |\partial_s^2\gamma|^2\eta_2 ds = 0. \tag{11}$$

If $\zeta(s,t) = \zeta_1(s,t)\partial_s\gamma(s,t) + \zeta_2(s,t)\partial_s^2\gamma(s,t) + \zeta_3(s,t)\partial_s\gamma(s,t) \times \partial_s^2\gamma(s,t)$satisfies $\langle\zeta, \eta\rangle = 0$, then we see that η_1 and η_2 satisfy Eqs. (10) and (11) and any η_3 satisfies

$$\int_0^L \{\zeta_1\eta_1 + \zeta_2\eta_2|\partial_s^2\gamma|^2 + \zeta_3\eta_3(\partial_s\gamma \times \partial_s^2\gamma)^2\}ds = 0. \tag{12}$$

In particular, if we assume $\eta_2 \equiv 0$ and $\eta_3 \equiv 0$, then we infer from Eq. (10) that $\eta_1 \equiv C$ holds true. Then, we deduce from Eq. (12) that

$$\int_0^L \zeta_1 ds = 0.$$

Now we define

$$\varphi(s,t) = \zeta_1(0,t) + \int_0^s \zeta_1(s,t)ds.$$

Then, φ has the period L and satisfies $\partial_s\varphi = \zeta_1$. Substituting it into Eq. (12) and using Eq. (10), we have

$$\begin{aligned}
0 &= \int_0^L \{\partial_s\varphi\ \eta_1 + \zeta_2\eta_2|\partial_s^2\gamma|^2 + \zeta_3\eta_3(\partial_s\gamma \times \partial_s^2\gamma)^2\}ds \\
&= \int_0^L \{-\varphi\partial_s\eta_1 + \zeta_2\eta_2|\partial_s^2\gamma|^2 + \zeta_3\eta_3(\partial_s\gamma \times \partial_s^2\gamma)^2\}ds \\
&= \int_0^L \{-\varphi\ |\partial_s^2\gamma|^2\eta_2 + \zeta_2\eta_2|\partial_s^2\gamma|^2 + \zeta_3\eta_3(\partial_s\gamma \times \partial_s^2\gamma)^2\}ds \\
&= \int_0^L \{(-\varphi + \zeta_2)\eta_2|\partial_s^2\gamma|^2 + \zeta_3\eta_3(\partial_s\gamma \times \partial_s^2\gamma)^2\}ds.
\end{aligned}$$

Recalling Eq. (11), we see that the vector-valued function (η_2, η_3) is orthogonal with $(|\partial_s^2\gamma|^2, 0)$ in the sense of L^2 inner product. Therefore, there exists some function $\mu = \mu(t)$ depending only on t such that

$$\left(\{-\varphi + \zeta_2\}\ |\partial_s^2\gamma|^2,\ (\partial_s\gamma \times \partial_s^2\gamma)^2\zeta_3\right) = \mu(|\partial_s^2\gamma|^2, 0)$$

that is,

$$-\varphi + \zeta_2 = \mu,\ \zeta_3 \equiv 0. \tag{13}$$

Setting

$$\mu + \varphi(s,t) = w(s,t),$$

the function $w(s, t)$ is the L periodic function and satisfies $\partial_s w = \zeta_1$. It follows from Eq. (13) that

http://dx.doi.org/10.5772/intechopen.69175

$$\zeta_2(s,t) = w(s,t).$$

Then $\zeta(s)$ is orthogonal with elements of V with respect to L^2 inner product. Thus, we get

$$\zeta(s,t) = \partial_s w(s,t)\partial_s\gamma(s,t) + w(s,t)\partial_s^2\gamma(s,t) = \partial_s\Big(w(s,t)\partial_s\gamma(s)\Big)$$

which completes the proof.

Q.E.D.

Remark 1. The assumption $\partial_s^2\gamma \neq \mathbf{0}$ in Lemma 1 means that the curvature is always non-zero. If $\partial_s^2\gamma = \mathbf{0}$ at some point, we cannot determine the tangential vector $\partial_s\gamma$ at the point uniquely. Therefore, we need the assumption in order to give a coordinate system at every point of Γ. On the other hand, in the case of a planar curve, we do not need the assumption. Indeed, by rotating the tangential vector, we can construct a coordinate system.

Remark 2. We mention the elastic flow with the inextensible condition (Eq. (3)), more precisely, L^2 gradient flow for $K(\gamma)$ under the constraint (Eq. (3)). To the best of our knowledge, the problem was first considered by N. Koiso [26] for planar closed curves. With the aid of smoothing effect of the elastic energy E, the Cauchy problem on the elastic flow has a unique classical solution and the solution converges to an equilibrium state as $t \to \infty$ in the C^∞-topology. The result can be extended to the following case: (i) L^2 gradient flow for E under the area-preserving condition and (C) [27] and (ii) L^2gradient flow for Tadjbakhsh-Odeh energy functional under the constraint (C) [28]. Moreover, the result [26] was also extended to the case of space curves [29].

2.4. Derivation of heat equation

In this subsection, we study the energy law. We confirm thermal energy conservation law (the first law of thermodynamics) and the increasing law of entropy (the second law of thermodynamics). To begin with, we consider the first law of thermodynamics. According to Ref. [30], thermal energy conservation law for thermoelastic system is given by

$$\theta\partial_t S + \nabla\cdot q = h \tag{14}$$

where S, q and h are entropy, thermal velocity and external heat, respectively. In our setting, thermal transfer occurs only on wire, and the wire does not expand. Then, we may regard $\nabla\cdot q$ as $\partial_s q$ as the same as one-dimensional case, where s is necessary to be arc length parameter. By the same reason, the Fourier law $q = \nabla\theta$ is replaced by

$$q = -\partial_s\theta. \tag{15}$$

The Helmholtz free energy density

$$\tilde{f} = \tilde{f}(\partial_s^2\gamma,\ \partial_s\gamma,\ \theta\,;\gamma^0)$$
$$= \frac{1}{2}|\partial_s^2\gamma\ |^2 + f(\partial_s\gamma,\ \theta\,;\gamma^0) + f_0(\theta)$$

and the entropy S are connected with the relation

$$S = -\frac{\partial\tilde{f}}{\partial\theta}.$$

Then, the conservation law, Eq. (14), is rewritten as follows:

$$-\theta f_0''(\theta)\,\partial_t\theta\ -\partial_s^2\theta = \theta\left(\partial_s\gamma^0\cdot\partial_s(\gamma-\gamma^0)\right)(\partial_t\partial_s\gamma\cdot\partial_s\gamma^0) + h. \tag{16}$$

We note that the Clausius-Duhem inequality holds automatically:

$$\partial_t S + \partial_s\left(\frac{q}{\theta}\right) \geq \frac{h}{\theta}\ .$$

Indeed, we observe from Eqs. (14) and (15) that

$$\partial_t S + \partial_s\left(\frac{q}{\theta}\right) = \frac{h-\partial_s q}{\theta} + \partial_s\left(\frac{q}{\theta}\right)$$
$$= \frac{h}{\theta} - \frac{q\partial_s\theta}{\theta^2}$$
$$= \frac{h}{\theta} + |\frac{\partial_s\theta}{\theta}|^2 \geq \frac{h}{\theta}\ .$$

The Clausius-Duhem inequality corresponds to the second law of thermodynamics. For more precise information of the inequality, we refer to, for example, 1.11 of chapter 4 in Ref. [2].

Here, we assume external heat source $h = 0$ and adopt the well-known form:

$$f_0(\theta) = \theta - \theta\log\theta.$$

Then since $f_0{}''(\theta) = -1/\theta,$ Eq. (16) is reduced to

$$\partial_t\theta - \partial_s^2\theta = \theta\left(\partial_s\gamma^0\cdot\partial_s(\gamma-\gamma^0)\right)(\partial_t\partial_s\gamma\cdot\partial_s\gamma^0).$$

We thus obtain the system of equation as Eq. (4).

3. Dual variation structure

Let us rewrite Eq. (4):

$$\partial_t^2\gamma + \partial_s^4\gamma\ + \partial_s f_{,\partial_s\gamma}(\partial_s\gamma,\ \theta;\partial_s\gamma^0) - \partial_s\{(v - 2\,|\partial_s^2\gamma|^2)\partial_s\gamma\} = 0, \tag{17}$$

$$-\partial_s^2 v + |\partial_s^2\gamma|^2 v = 2|\partial_s^2\gamma|^4 - |\partial_s^3\gamma|^2 + |\partial_s\partial_t\gamma|^2 + \partial_s^2 f_{,\partial_s\gamma}(\partial_s\gamma,\ \theta;\partial_s\gamma^0)\cdot\partial_s\gamma, \tag{18}$$

$$\partial_t\theta - \partial_s^2\theta = \theta\Big(\partial_s\gamma^0 \cdot \partial_s(\gamma - \gamma^0)\Big)(\partial_t\partial_s\gamma \cdot \partial_s\gamma^0), \ (t, s) \in (0, T) \times S_L^1, \tag{19}$$

$$\gamma(0, s) = \gamma_0(s), \ \partial_t\gamma(0, s) = \gamma_1, \ \theta(0, s) = \theta_0, \ s \in S_L^1. \tag{20}$$

In this section, we show that the problem in Eqs. (17)–(20) has also dual variation structure as well as the problem (Eq. (2)). The structure plays an important role to prove the dynamical stability of infinitesimally stable stationary state.

We assume that the system has sufficiently smooth solution (γ, θ) satisfying $\theta > 0$. Then initial data also has to satisfy

$$|\partial_s\gamma_0| = 1, \ \partial_s\gamma_0 \cdot \partial_s\gamma_1 = 0.$$

Setting

$$\begin{aligned} f_1(\partial_s\gamma) &:= \frac{1}{2}\left(\frac{\partial\gamma^0}{\partial s} \cdot \frac{\partial(\gamma - \gamma^0)}{\partial s}\right)^2, \\ f_2(\partial_s\gamma) &:= \frac{1}{6}\left(\frac{\partial\gamma^0}{\partial s} \cdot \frac{\partial(\gamma - \gamma^0)}{\partial s}\right)^6 - \frac{1}{4}\left(\frac{\partial\gamma^0}{\partial s} \cdot \frac{\partial(\gamma - \gamma^0)}{\partial s}\right)^4 - \frac{\theta_c}{2}\left(\frac{\partial\gamma^0}{\partial s} \cdot \frac{\partial(\gamma - \gamma^0)}{\partial s}\right)^2, \end{aligned}$$

we have the relation $f = \theta f_1 + f_2$. Multiplying Eq. (17) by $\partial_t\gamma$ and integrating it with respect to s, we obtain

$$\frac{d}{dt}\left(\frac{1}{2}||\partial_t\gamma||_{L^2}^2 + \frac{1}{2}||\partial_s^2\gamma||_{L^2}^2\right) = -\langle f_{,\partial_s\gamma}, \ \partial_s\partial_t\gamma\rangle = -\frac{d}{dt}\int_0^L f_2(\partial_s\gamma)ds - \int_0^L \theta\partial_t f_1(\partial_s\gamma)ds.$$

Integrating Eq. (19), we find

$$\frac{d}{dt}\int_0^L \theta ds = \int_0^L \theta\partial_t f_1(\partial_s\gamma)ds.$$

Then for the quantity

$$E(\gamma, \partial_t\gamma, \theta) := \frac{1}{2}||\partial_t\gamma||_{L^2}^2 + \frac{1}{2}||\partial_s^2\gamma||_{L^2}^2 + \int_0^L \theta ds + \int_0^L f_2(\partial_s\gamma)ds,$$

it holds that

$$\frac{d}{dt}E(\gamma, \partial_t\gamma, \theta) = 0.$$

Moreover, for the quantity

$$W(\partial_s\gamma, \theta) := \int_0^L \{f_1(\partial_s\gamma) - \log\theta\}ds$$

we deduce from Eq. (19) that

$$\frac{d}{dt}W(\partial_s\gamma,\theta)=\int_0^L\frac{\partial_s^2\theta}{\theta}ds=-\int_0^L|\frac{\partial_s\theta}{\theta}|^2ds\leq 0.$$

Finally, we confirm an important structure of the system (17)–(20). We denote the stationary state of θ by $\overline{\theta}>0$, and the corresponding equilibrium by γ satisfy the following constraint:

$$b\equiv E(\gamma_0,\ \gamma_1,\ \theta_0)=E(\gamma,\ 0,\ \overline{\theta}),$$

$$\partial_s^4\gamma=\partial_s\{\overline{\theta}f_{1,\partial_s\gamma}+f_{2,\partial_s\gamma}\}.$$

Eliminating $\overline{\theta}$ by the relation

$$b=L\overline{\theta}+\frac{1}{2}||\partial_s^2\gamma||_{L^2}^2+\int_0^L f_2(\partial_s\gamma)\ ds$$

the stationary state of this problem satisfies the following nonlinear nonlocal problem:

$$\partial_s^4\gamma=\partial_s\left\{\frac{1}{L}\left(b-\frac{1}{2}||\partial_s^2\gamma||_{L^2}^2-\int_0^L f_2(\partial_s\gamma)ds\right)f_{1,\partial_s\gamma}(\partial_s\gamma)+f_{2,\partial_s\gamma}(\partial_s\gamma)\right\}. \tag{21}$$

Eq. (21) is derived as the Euler-Lagrange equation of the functional

$$J_b(y):=\frac{1}{L}\int_0^L f_1(y)ds-\log\left(b-\frac{1}{2}||\partial_s y||_{L^2}^2-\int_0^L f_2(y)ds\right)+\log L$$

where $y=\ \partial_s\gamma\in H^2(S_L^1,S^2)$ and

$$S^2:=\{\omega\in\mathbf{R}^3||\omega|=1\}.$$

We remark that the following relation between J_b and W holds true:

$$W(\partial_s\gamma,\theta)\geq LJ_b(\partial_s\gamma).$$

The relation is called **semi-unfolding minimality**. Thus, if $(\gamma,\ \theta)$ is a non-stationary state, $b=E(\gamma_0,\gamma_1,\theta_0)$, $\overline{y}=\partial_s\overline{\gamma}$ is a linearized stable critical point of $J_b=J_b(y)$, $y\in\ H^1(S_L^1,\ S^2)$ and $\overline{\theta}>\ 0$ is a constant satisfying $E(\overline{\gamma},0,\overline{\theta})\ =\ b$, then it holds that for $y=\partial_s\gamma$

$$J_b(y)-J_b(\overline{y})\leq W(\partial_s\gamma_0,\ \theta_0)-W(\overline{y},\overline{\theta}).$$

By this structure, we can infer that any infinitesimally stable stationary state is dynamically stable, that is, stable in the Lyapunov sense. A critical point $\overline{y}=\partial_s\overline{\gamma}$ of J_b for $\overline{\gamma}\in H^2(S_L^1,S^2)$ is **infinitesimally stable** if there exists $\varepsilon_0>0$ such that any $\varepsilon_1\in\left(0,\frac{\varepsilon_0}{2}\right]$ admits $\delta_0>0$ such that if $||\partial_s(\gamma-\overline{\gamma})||_{H^1}<\varepsilon_0$ and $J_b(\partial_s\gamma)-\ J_b(\partial_s\overline{\gamma})<\delta_0$ then

$$||\partial_s(\gamma - \overline{\gamma})||_{H^1} < \varepsilon_1.$$

The definition of infinitesimally stable is obviously weaker than the one of well-known linearized stable which means that for a critical point $\overline{y} = \partial_s\overline{\gamma}$ of J_b, the quadratic form

$$Q_y(w, w) = \frac{d}{d\varepsilon^2} J_b(\overline{y} + \varepsilon w)|_{\varepsilon=0}$$

is a positive definite for any $w \in H^2(S_L^1, S^2)$.

Theorem 1. *Assume that $\overline{\theta} > 0$ is a constant and that $\overline{\gamma}$ is an infinitesimally stable critical point of J_b with constraint $\partial_s\gamma \in H^2(S_L^1, S^2)$. Then $(\overline{\gamma}, \overline{\theta})$ is a dynamically stable in the sense that for any $\varepsilon > 0$, there exists $\delta > 0$ such that if*

$$E(\gamma_0, \gamma_1, \theta_0) = b, ||\partial_s(\gamma_0 - \overline{\gamma})||_{H^1} < \delta, \ |\frac{1}{L}\int_0^L \log\theta_0(s)ds - \log\overline{\theta}| < \delta \tag{22}$$

then

$$\sup_{t\geq 0}||\partial_s\left(\gamma(t) - \overline{\gamma}\right)||_{H^1} < \varepsilon, \ |\frac{1}{L}\int_0^L \log\theta(t, s)ds - \log\overline{\theta}| < \varepsilon.$$

Proof. We first show the semi-unfolding minimality. From the energy conservation law, we see that

$$b = \frac{1}{2}||\partial_t\gamma||_{L^2}^2 + \frac{1}{2}||\partial_s^2\gamma||_{L^2}^2 + \int_0^L \theta ds + \int_0^L f_2(\partial_s\gamma)ds.$$

It follows from the Jensen inequality that

$$\frac{1}{L}\int_0^L \log\theta ds \leq \log\left(\frac{1}{L}\int_0^L \theta ds\right).$$

Then we have

$$\begin{aligned} W(\partial_s\gamma, \theta) &\geq \int_0^L f_1(\partial_s\gamma)ds - \log\left(\frac{1}{L}\int_0^L \theta ds\right) \\ &\geq \int_0^L f_1(\partial_s\gamma)ds - L\log\frac{1}{L}\left(b - \frac{1}{2}||\partial_s y||_{L^2}^2 - \int_0^L f_2(y)ds\right) \\ &= LJ_b(\partial_s\gamma). \end{aligned}$$

We have thus completed to show the semi-unfolding minimality. Recall that $\overline{\gamma} \in H^2(S_L^1, S^2)$ is an infinitesimally stable critical point of $J_b(\partial_s\gamma)$. Thus, we find $\varepsilon_0 > 0$such that any $\varepsilon_1 \in (0, \frac{\varepsilon_0}{2}]$ admits $\delta_0 > 0$ such that if $||\partial_s(\gamma - \overline{\gamma})||_{H^1} < \varepsilon_0$ and $J_b(\partial_s\gamma) - J_b(\partial_s\overline{\gamma}) < \delta_0$ then

$$||\partial_s(\gamma - \overline{\gamma})||_{H^1} < \varepsilon_1. \tag{23}$$

From the above properties, it holds that

$$J_b\left(\partial_s\gamma(t)\right) \leq \frac{1}{L}W\left(\partial_s\gamma(t),\ \theta(t)\right) \leq \frac{1}{L}W(\partial_s\gamma_0,\ \theta_0). \tag{24}$$

Moreover, for the constant $\overline{\theta} > 0,$ we obtain

$$W(\partial_s\overline{\gamma},\ \overline{\theta}) = L\left[\frac{1}{L}\int_0^L f_1(\partial_s\gamma)ds\ - \log\frac{1}{L}\left(b - \frac{1}{2}||\partial_s\overline{y}||_{L^2}^2 - \int_0^L f_2(\overline{y})ds\right)\right] = LJ_b(\partial_s\,\overline{\gamma}),$$

namely,

$$J_b(\partial_s\overline{\gamma}) = \frac{1}{L}W(\partial_s\overline{\gamma}, \overline{\theta}). \tag{25}$$

Given $\varepsilon > 0$, setting $\delta \in \left(0,\ \frac{\varepsilon_0}{2}\right]$ and satisfying Eq. (22), we have

$$\begin{aligned}\frac{1}{L}|W(\partial_s\gamma_0,\ \theta_0) - W(\partial_s\overline{\gamma}, \overline{\theta})| \leq \frac{1}{L}\left(||f_{1,\ \partial_s\gamma}(\partial_s\gamma_0)||_{L^\infty} + ||f_{1,\ \partial_s\gamma}(\partial_s\overline{\gamma})||_{L^\infty}\right)||\partial_s(\gamma_0 - \overline{\gamma})||_{L^1} \\ +|\frac{1}{L}\int_0^L \log\theta_0 ds - \log\overline{\theta}| < \min(\delta_0, \varepsilon).\end{aligned} \tag{26}$$

Therefore, it follows from Eq. (24) to Eq. (26) that

$$J_b\left(\partial_s\gamma(t)\right) - J_b(\partial_s\overline{\gamma}) = \frac{1}{L}\left(W(\partial_s\gamma_0,\ \theta_0) - W(\partial_s\overline{\gamma}, \overline{\theta})\right) < \delta_0.$$

If $||\partial_s(\gamma(t) - \overline{\gamma})||_{H^1} = \delta(\leq \varepsilon_0/2\ < \varepsilon_0)$, then we apply Eq. (23) for $\varepsilon_1 = \delta$, and hence

$$||\partial_s\left(\gamma(t) - \overline{\gamma}\right)||_{H^1} < \delta,$$

which is a contradiction. Thus, we have

$$||\partial_s\left(\gamma(t) - \overline{\gamma}\right)||_{H^1} \neq \delta.$$

Here, from $\gamma \in C\left([0, \infty); H^2\right)$ and $||\partial_s(\gamma_0 - \overline{\gamma})||_{H^1} < \delta,$ it follows that

$$||\partial_s\left(\gamma(t) - \overline{\gamma}\right)||_{H^1} < \delta \tag{27}$$

for any $t \geq 0$.

From the semi-unfolding minimality (Eqs. (24) and (25)) and the linearized stability of J_b, we observe that

$$W(\partial_s\gamma,\theta) \geq LJ_b(\partial_s\gamma) \geq LJ_b(\partial_s\overline{\gamma}) = W(\partial_s\overline{\gamma},\overline{\theta}).$$

Then, combining Eq. (26) with Eq. (27), we have

$$\left|\frac{1}{L}\int_0^L \log\theta(t,s)ds - \log\overline{\theta}\right| \leq \frac{1}{L}\left(W(\partial_s\gamma,\theta) - W(\partial_s\overline{\gamma},\overline{\theta})\right) + \frac{1}{L}\left|\int_0^L \{f_1(\partial_s\gamma) - f_1(\partial_s\overline{\gamma})\}ds\right|$$
$$\leq \frac{1}{L}\left(W(\partial_s\gamma_0,\theta_0) - W(\partial_s\overline{\gamma},\overline{\theta})\right) + C||\partial_s(\gamma-\overline{\gamma})||_{H^1} \leq \varepsilon + C\delta \leq 2\varepsilon$$

where δ is small enough such that $\delta < \varepsilon / C$. This completes the proof.

Q.E.D.

Remark 3. Both the existence of solution for evolution equations (Eq. (4)) and non-trivial solutions for stationary problem (Eq. (21)) are open problems. In the straight material case (i.e. the problem (2)), smooth solution for Eq. (2) is assured in [11] (we also refer to chapter 5 in [2]). The existence results of non-trivial solution for stationary problem (Eq. (21)) in low-temperature and low-energy cases can be found in [21, 31].

4. Concluding remarks

In this chapter, we propose the new mathematical model describing the movement of wire made of shape memory alloys. The derived system of nonlinear partial differential equations is a thermoelastic system with phase transition and non-stretching constraint. The Falk model (Eq. (2)) represents the dynamics for crystal as a stack of layers, whose displacement is restricted to move only on one direction. On the other hand, our model describes the dynamics of wire. We emphasize that our model allows the displacement of each direction. Thus, our model may describe a more realistic motion of wire made of shape memory alloys. Moreover, it is also interesting to regard our model as a mathematical problem on elastic curve with heat conduction. To the best of our knowledge, there is no result considering such a mathematical problem.

We mention the mathematical contribution of the present chapter. We prove the dynamical stability of an infinitesimally stable stationary state by finding the dual variation structure in our model. This property shall be applicable, for example, to assure the strength of not only a wire in an original shape but also a deformed wire. Indeed, in the straight material case, namely in the Falk model (Eq. (2)), numerical simulation shows the stability in this sense (see [17]).

Acknowledgements

This work was partially supported by JSPS KAKENHI Grant-in-Aid for Scientific Research (A) and (C), Grant Numbers 26247013 and 16K05234.

Author details

Shinya Okabe[1], Takashi Suzuki[2]* and Shuji Yoshikawa[3]

*Address all correspondence to: suzuki@sigmath.es.osaka-u.ac.jp

1 Mathematical Institute, Tohoku University, Sendai, Japan

2 Graduate School of Engineering Science, Osaka University, Osaka, Japan

3 Faculty of Science and Technology, Oita University, Oita, Japan

References

[1] Falk F. Elastic phase transitions and nonconvex energy functions. In: Pitman Research Notes in Mathematics Series. Vol 185. IRSEE. Harlow: Longman Science and Technology; 1990. pp. 45–59

[2] Brokate M, Sprekels J. Hysteresis and phase transitions (Applied Mathematical Sciences). 121st ed. New York: Springer-Verlag; 1996

[3] Jiang S, Racke R. Evolution Equations in Thermoelasticity. Monographs and Surveys in Pure and Applied Mathematics. 112nd ed. Boca Raton: Chapman & Hall/CRC; 2000

[4] Fremond M. Non-Smooth Thermomechanics. Berlin, Heidelberg: Springer-Verlag; 2002. p. 480. DOI: 10.1007/978-3-662-04800-9

[5] Fremond M, Miyazaki S. Shape Memory Alloys. Wien: Springer-Verlag; 1996. p. 147. DOI: 10.1007/978-3-7091-4348-3

[6] Fang Y-F, Grillakis MG. Existence and uniqueness for Boussinesq type equations on a circle. Communications in Partial Differential Equations. 1996;**21**:1253–1277

[7] Falk F, Laedke EW, Spatschek KH. Stability of solitary-wave pulses in shape-memory alloys. Physical Review B. 1987;**36**:3031–3041

[8] Friedman A, Sprekels J. Steady states of austenitic-martensitic domains in the Ginzburg-Landau theory of shape memory alloys. Continuum Mechanics and Thermodynamics 1990;**2**:199–213

[9] Garcke H. Travelling wave solutions as dynamic phase transitions in shape memory alloys. Journal of Differential Equations. 1995;**121**:203–231

[10] Tsutsumi Y, Yoshikawa S. Invariant measures for the isothermal Falk model of shape memory alloys. GAKUTO International Series. Mathematical Sciences and Applications. 2015;**37**:163–182

[11] Sprekels J, Zheng S. Global solutions to the equations of a Ginzburg-Landau theory for structural phase transitions in shape memory alloys. Journal of Physics D. 1989;**39**:59–76

[12] Hoffmann K-H, Zochowski A. Analysis of the thermoelastic model of a plate with nonlinear shape memory alloy reinforcements. Mathematical Methods in the Applied Sciences. 1992;**15**:631–645

[13] Yoshikawa S. Weak solutions for the Falk model system of shape memory alloys in energy class. Mathematical Methods in the Applied Sciences. 2005;**28**:1423–1443

[14] Hoffmann K-H, Zou J. Finite element approximations of Landau-Ginzburg's equation model for structural phase transitions in shape memory alloys. RAIRO Mathematical Modelling and Numerical Analysis. 1995;**29**:629–655

[15] Niezgodka M, Sprekels J. Convergent numerical approximations of the thermomechanical phase transitions in shape memory alloys. Numerische Mathematik. 1991;**58**:759–778

[16] Matus P, Melnik RVN, Wang L, Rybak I. Applications of fully conservative schemes in nonlinear thermoelasticity: Modelling shape memory materials. Mathematics and Computers in Simulation. 2004;**65**:489–509

[17] Yoshikawa S. A conservative finite difference scheme for the Falk model system of shape memory alloys. ZAMM Zeitschrift für Angewandte Mathematik und Mechanik. 2015;**95** (12):1393–1410

[18] Yoshikawa S. An error estimate for structure-preserving finite difference scheme for the Falk model system of shape memory alloys. IMA Journal of Numerical Analysis. 2017;**37**:477–504

[19] Suzuki T, Yoshikawa S. Stability of the steady state for the Falk model system of shape memory alloys. Mathematical Methods in the Applied Sciences. 2007;**30**:2233–2245

[20] Suzuki T. Mean Field Theories and Dual Variation—Mathematical Structures of the Mesoscopic Model. 2nd ed. Amsterdam, Paris: Atlantis Press; 2016

[21] Suzuki T, Tasaki S. Stationary solution to the Falk system on shape memory alloys. Mathematical Methods in the Applied Sciences. 2010;**33**:994–1011

[22] Suzuki T, Yoshikawa S. Stability of the steady state for multi-dimensional thermoelastic systems of shape memory alloys. Discrete and Continuous Dynamical Systems—Series S. 2012;**5**:209–217

[23] Gage M. Curve shortening makes convex curves circular. Inventiones Mathematicae. 1984;**76**:357–364

[24] Gage M, Hamilton R. The shrinking of convex plane curves by the heat equation. Journal of Differential Geometry. 1986;**23**:69–96

[25] Grayson MA. The heat equation shrinks embedded plane curves to round points. Journal of Differential Geometry. 1987;**26**:285–314

[26] Koiso N. Motion of a springy circle wire. Science Reports—College of General Education, Osaka University. 1991;**40**:15–19

[27] Okabe S. The motion of elastic planar closed curves under the area-preserving condition. Indiana University Mathematics Journal. 2007;**56**:1871–1912

[28] Okabe S. The dynamics of elastic closed curves under uniform high pressure. Calculus of Variations and Partial Differential Equations. 2008;**33**:493–521

[29] Koiso N. On motion of a curve towards elastica. In: Actes da la Table Ronde de Geometrie Differentielle, Semin Congr. Vol. 1, 1992; Luminy. Paris: Soc. Math. France; 1996. pp. 403–436

[30] Lagnese J, Lions J-L. Modelling Analysis and Control of Thin Plates. Recherches en Mathematiques Appliquees (Research in Applied Mathematics). 6th ed. Paris, Masson; 1988

[31] Suzuki T, Tasaki S. Stationary solutions to a thermoelastic system on shape memory materials. Nonlinearity. 2010;**23**:2023–2656

Section 2

Shape Memory Alloys: Applications

Chapter 3

Magnetocaloric Effects in Metamagnetic Shape Memory Alloys

Takumi Kihara, Xiao Xu, Wataru Ito,
Ryosuke Kainuma, Yoshiya Adachi,
Takeshi Kanomata and Masashi Tokunaga

Additional information is available at the end of the chapter

http://dx.doi.org/10.5772/intechopen.69116

Abstract

Recently, metamagnetic shape memory alloys have attracted much attention as candidates for the rare-earth free magnetic refrigerants. These materials undergo the martensitic transformation (MT) at around room temperature accompanied by a significant entropy change. The application of the magnetic field at the low-temperature martensitic phase realizes the magnetic field-induced martensitic transformation (MFIMT). Through the MFIMT, the materials show an unconventional magnetocaloric effect (MCE), which is called inverse magnetocaloric effect (IMCE). In this chapter, the direct measurement system of MCE in pulsed-high-magnetic fields is introduced. With taking the advantage of the fast field-sweep rate of pulsed field, adiabatic measurements of MCE are carried out at various temperatures. Using this technique, the IMCEs of the metamagnetic shape memory alloys NiCoMnIn and NiCoMnGa are directly measured as adiabatic temperature changes in pulsed fields. From the experimental data of MCE for NiCoMnIn, the entropy of spin system in the austenite phase is estimated through a simple mean-field model. By the combination of MCE, magnetization and specific heat measurements, the electronic, lattice and magnetic contributions to the IMCE are individually evaluated. The result for NiCoMnIn demonstrates that lattice entropy plays the dominant role for IMCE in this material.

Keywords: shape memory alloys, Heusler alloys, magnetocaloric effects, specific heat, pulsed magnetic field

INTECH
open science | open minds

1. Introduction

1.1. Magnetocaloric effects and magnetic refrigeration

Magnetocaloric effect (MCE) is thermodynamically defined as a temperature change of magnetic material due to the variation of an externally applied magnetic field. This effect is a consequence of the field variation of the entropy of a material. For instance, the application of a magnetic field to a paramagnet in the adiabatic condition reduces the disorder of spins, which lowers the magnetic entropy in the material. Since the total entropy is conserved in an adiabatic condition, the lattice entropy (and the electronic entropy in case of metals) must increase by a comparable amount of the magnetic entropy decrease. Consequently, the temperature of the material increases. In the adiabatic demagnetization process, the magnetic entropy is restored from the lattice and electronic systems, and hence, the temperature of the material decreases. In the MCE measurement, the adiabatic temperature change or the isothermal entropy change as a function of applied field is the experimentally measurable quantity. Since the MCE is sensitive to the entropy change of the material at the magnetic phase transitions, it is useful to map out the magnetic phase diagram and evaluate the temperature and field variations of entropy by the combination with the specific heat measurement. The MCE is also applied to the magnetic refrigeration, which is realized by means of the cycle of adiabatic magnetization/demagnetization and heat exchange between the magnetic material and the surrounding. This technique has long been used to realize extremely low temperatures [1, 2].

In recent decades, the magnetic refrigeration based on the MCE has attracted much attention as an alternative technique to the vapor-compression cycle for the ambient temperatures, which has been triggered by the discovery of the giant MCE (GMCE) at around room temperatures [3–5]. The GMCE is first observed in $Gd_5Si_2Ge_2$, which was reported by Pecharsky and Gschneidner [3]. This material undergoes the first-order magnetic and structural phase transition at 276 K with a giant entropy change. The maximum entropy change at the transition temperature in this material is about twice larger than that of pure Gd, which is considered as a benchmark of magnetocaloric material at around room temperature. The GMCE, in this material, appears as a result of the simultaneous changes in the magnetic and lattice entropy through the first-order phase transition (FOT). Hence, the total entropy change can exceed the limit of the magnetic entropy change caused by the spin ordering. For the simultaneous change in the magnetic and lattice entropy, the strong coupling between the spin and the lattice systems is needed. For this reason, the GMCE undergoes in many cases the first-order magneto-structural phase transition [3–12].

1.2. MCEs in Ni-Mn-based Heusler alloys

$Ni_{50}Mn_{50-x}Z_x$ (Z = Ga, In, Sn and Sb) Heusler alloys have attracted a lot of attention because of the potential applications for novel rare-earth free magnetic refrigerants [13–17]. These alloys have a cubic ($L2_1$) Heusler structure with a space group of $Fm\bar{3}m$ [15]. Some of the non-stoichiometric compositions undergo martensitic transformation (MT) from a high-temperature austenitic phase (A-phase: cubic) to a low-temperature martensitic phase of reduced symmetry (M-phase: tetragonal, orthorhombic, or monoclinic) at around room temperature ($Ni_{50}Mn_{25}Ga_{25}$ is the only Ni-Mn-based Heusler alloy that shows the MT in the stoichiometric composition [18]). In these alloys, the MT temperatures (T_{MT}) and the magnetic properties in each phase strongly depend on the Z species and its compositional ratio x [15]. The T_{MT} increases with decreasing x. In the small x region, the ferromagnetic phase appears at the Curie temperatures in both the A phase (T_C^A) and the M phase (T_C^M), respectively [15, 16, 19–21]. In addition, the partial substitution of Ni with Co enhances the exchange coupling between the spins in the alloys, which contributes to increase T_C^A and decrease T_C^M and T_{MT} [14, 22]. $Ni_{45}Co_5Mn_{36.7}In_{13.3}$ undergoes MT accompanied with a magnetic phase transition from a ferromagnetic A phase to a paramagnetic M phase (monoclinic). The application of an external magnetic field to the M phase realizes magnetic-field-induced martensitic transformation (MFIMT). Kainuma et al. [14] reported, for the first time, the magnetic-field-induced shape recovery by MFIMT in this material, which is called metamagnetic shape memory effect. Owing to the large magnetic-field-induced strain and the large output stress through MFIMT, $Ni_{45}Co_5Mn_{36.7}In_{13.3}$ is considered to be a potential magnetic actuator material [14].

In the $Ni_{45}Co_5Mn_{36.7}In_{13.3}$, the MFIMT occurs as it is accompanied by significant increase of entropy, whereas the applied magnetic field aligns the spins parallel to the field direction and reduces the magnetic entropy [14, 22]. This unconventional phenomenon is called the inverse magnetocaloric effect (IMCE). The IMCEs were also observed in other compositions $Ni_{50}Mn_{50-x}Z_x$ (Z = Ga, In, Sn and Sb) [13–16, 19–23]. The IMCE indicates the significant positive change in entropy through the MFIMT that exceeds the negative contribution from the spin ordering. Therefore, the individual evaluations of the electronic, lattice and magnetic entropy changes are important to elucidate the origin of such positive entropy changes at the MFIMT. In addition, the direct measurement of MCE is also important because it enables us to evaluate the magnetic part of entropy as presented in Section 4. However, numerous studies have attempted to understand the IMCE in these materials through indirect methods such as magnetization and/or specific heat, which can only reveal the total entropy change [13–17, 19–23].

There are several studies investigating adiabatic temperature change of these Heusler alloys under magnetic fields. For instance, Liu et al. [17] carried out the direct measurement of IMCEs up to 1.9 T for NiMnInCo. In their experiments, the samples undergo the MFIMT from a pure M phase to a mixed phase because the magnetic field of 1.9 T is insufficient to complete the MFIMT. In that case, the quantitative interpretation of the MCE results is difficult. Therefore, the direct MCE measurements in the wide range of temperatures and magnetic fields are crucial to provide a greater understanding of this phenomenon.

2. Thermodynamics of MCE

In this section, a brief explanation of thermodynamics of the MCE is provided (refer to Ref. [24] for more details). Let us start with the total entropy of the system $S(T, H, p)$. The total differential of $S(T, H, p)$ can be written as:

$$dS = \left(\frac{\partial S}{\partial T}\right)_{H,p} dT + \left(\frac{\partial S}{\partial H}\right)_{T,p} dH + \left(\frac{\partial S}{\partial p}\right)_{T,H} dp \tag{1}$$

Here, T, H and p denote temperature, external magnetic field and pressure, respectively. In the adiabatic and isobaric condition ($dS = 0$ and $dp = 0$), one obtains:

$$dT = -\left(\frac{\partial T}{\partial S}\right)_H \left(\frac{\partial S}{\partial H}\right)_T dH. \tag{2}$$

Since the heat capacity of a system at a constant magnetic field $C_H(T)$ is defined as $C_H(T) = T(\partial S / \partial T)_H$, the isentropic temperature change due to the variation of the magnetic field from H_1 to H_2 can be expressed as:

$$\Delta T_{ad}(H_1 \rightarrow H_2) = -\int_{H_1}^{H_2} \frac{T}{C_H(T)} \left(\frac{\partial S}{\partial H}\right)_H dH. \tag{3}$$

On the other hand, the isothermal-isobaric entropy change is given by the well-known Maxwell relation:

$$\left(\frac{\partial S}{\partial H}\right)_{T,p} = \left(\frac{\partial M}{\partial T}\right)_{H,p} \tag{4}$$

Here, M is the magnetization of a material. After integration, Eq. (4) gives

$$\Delta S_T(H_1 \rightarrow H_2) = -\int_{H_1}^{H_2} \left(\frac{\partial M(T,H)}{\partial T}\right)_H dH. \tag{5}$$

By combining Eqs. (3) and (4), the ΔT_{ad} can be expressed as:

$$\Delta T_{ad}(H_1 \rightarrow H_2) = -\int_{H_1}^{H_2} \frac{T}{C_H(T)} \left(\frac{\partial M}{\partial T}\right)_H dH. \tag{6}$$

Using Eqs. (5) and (6), ΔT_{ad} and ΔS_T can indirectly be estimated from the magnetization and specific heat measurements. Therefore, many studies employed so far use this method to evaluate the magnetocaloric properties for the several materials [3–16].

When a material undergoes magnetic phase transitions, the MCEs at the phase boundary show different behaviour between first- and second-order phase transitions. In the case of the second-order phase transition, the entropy of a material continuously and reversibly changes, and hence, Eqs. (5) and (6) can safely be used. On the other hand, the entropy discontinuously changes through an FOT, and the heat capacity becomes infinite at the transition temperature. Since the Maxwell relation is relevant only when the entropy is a continuous function of temperature and magnetic field, the direct application of the Eqs. (5) and (6) for the FOT is not allowed. Recently, several works proposed some approaches to apply the Maxwell relations to the FOTs, but it is still in debate [25–31]. Therefore, the direct measurement of MCE (ΔT_{ad}) is crucial to gain a deep insight into the entropic behaviour through the FOTs.

3. Direct MCE measurement system in pulsed-high-magnetic field

As described in the previous sections, the MCE measurements are of particular interest in both fundamental (investigation of the magnetic phase transitions) and applied physics (magnetic refrigeration). Therefore, a lot of experimental techniques of the MCE measurements have been developed so far [32–37]. Recently, Kihara et al. [36] developed a direct measurement system of MCE under pulsed high-magnetic fields up to 56 T. In this section, the details of their experimental technique are provided.

Figure 1(a) shows the schematic view of the electric circuit for the pulsed field generation system. The magnetic fields are generated in the solenoid magnet coil by discharging the energy stored in the capacitor bank [38]. **Figure 1(b)** shows the time variations of the pulsed fields generated in the nondestructive magnet installed at the Institute for Solid State Physics (ISSP), the University of Tokyo. The three curves correspond to the field profiles for charging voltages of 1, 5 and 8 kV to the capacitor bank, respectively. As shown in **Figure 1(b)**, the maximum field depends on the charging voltage to the capacitor bank. The total duration of the pulsed field is about 36 ms. This fast sweep rate of the pulsed field (the maximum sweep rate is about 10^4 T/s) has an advantage to realize the adiabatic conditions.

Figures 2(a) and **(b)** show the schematic and the picture of the sample setup. The sample is shaped into the thin plate with the thickness of less than 0.1 mm to reduce the eddy current heating caused by the application of the pulsed field (in case of metallic samples). The mag-

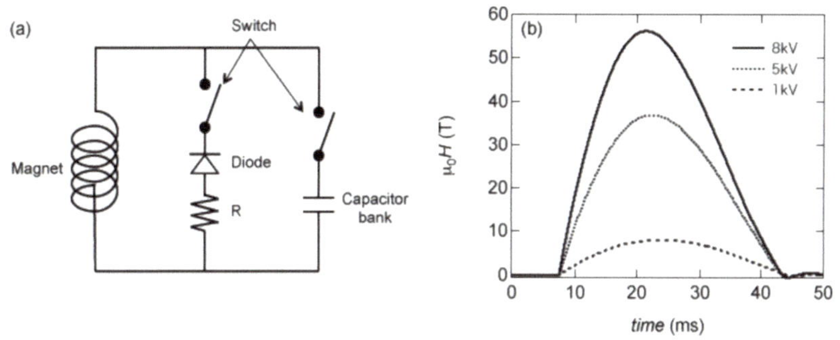

Figure 1. (a) A schematic view of the electric circuit of the pulse magnet system and (b) Magnetic field profiles of the 56-T pulse magnet at the Institute for Solid State Physics, The University of Tokyo.

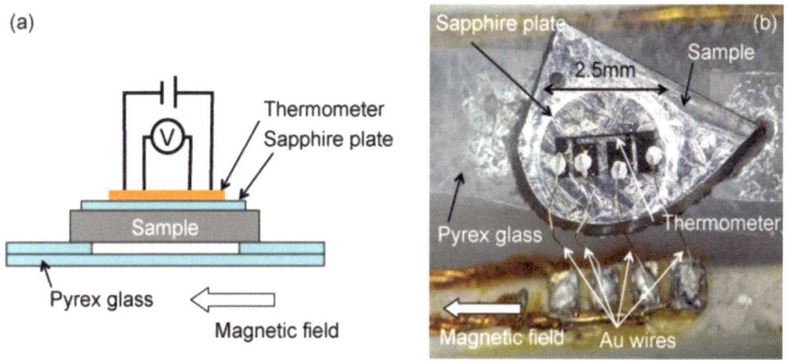

Figure 2. (a) A schematic drawing of the sample with the thermometer and (b) A picture of the sample ($Ni_{41}Co_9Mn_{31.5}Ga_{18.5}$) mounted on the MCE probe.

netic field is applied parallel to the sample plane to reduce the cross-section of the sample. In the thermometry under pulsed fields, the sample temperature changes rapidly as a function of the magnetic field. Therefore, commercial thermometers such as a Cernox® bare chip cannot be used because of the limitation of their thermal response time [35]. To detect the instantaneous change in the sample temperature in the pulsed field, a very small resistive thermometer is used, in which a patterned Au sensor (thickness: 100 nm) is deposited on a sapphire disk (thickness: 5 μm). This thermometer is mounted on top of the sample with a small amount of Apiezon® N grease as shown in **Figure 2(b)**. The small heat capacity of this thermometer (typical value is about 200 μJ/K at 300 K) and the large thermal conductance of the sapphire disk realize the fast response to the sample temperature and enable us to measure the MCEs in the pulsed fields as demonstrated for the NiCoMnIn in the next section. The sample with the thermometer is fixed on the Pyrex® glass plates, which have low thermal conductance, by the small amount of glue in order to reduce the heat leak to the surrounding. The probe (assembly of the sample, the thermometer, the sample holder, the bath heater, etc.) is inserted in the thin-walled tube made of non-magnetic stainless steel as shown in **Figure 3**. The sample space (inside of the tube) is evacuated to reduce the heat exchange between the sample and the surrounding through the residual gas. The tube is immersed in liquid helium for low temperature measurements. For high temperature measurements, small amount of helium gas is introduced into the space indicated by "Liq. He" in **Figure 3** as a heat exchange gas. During the measurement, the probe is cooled through the cold finger that is connected to the tube at the bottom of the probe. The sample temperature is regulated by the feedback operation of the bath heater (**Figure 3**). As shown in **Figure 2(b)**, the four Au wires (diameter: 30 μm) are connected to the thermometer by the small amount of Ag paste. The resistance of the thermometer is measured by the ac method using numerical lock-in technique [35, 36].

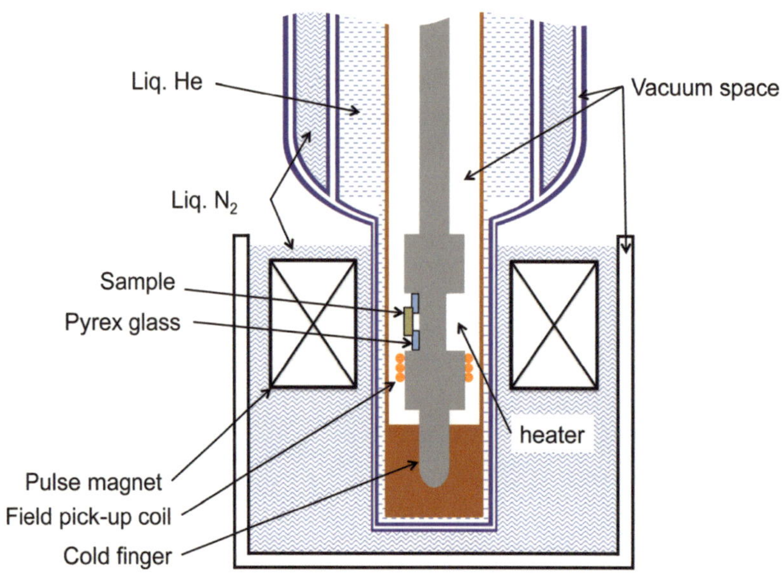

Figure 3. A schematic drawing of the probe setup.

To calibrate the Au thermometer grown on the sapphire disk, the longitudinal magnetoresistances (MRs) are measured in the pulsed fields before the MCE measurements. To carry out the isothermal measurements, the sample is removed, and the sapphire disk with the thermometer is placed directly on the Pyrex® glass substrate. Moreover, 1 atm of helium gas is introduced into the sample space at room temperature. **Figure 4** shows the MRs of the thermometer measured up to 56 T at the various temperatures (black curves). At 0 T, one can confirm the linear temperature dependence of the resistance of the metallic Au film. The slope is dR / dT = 8.97 mΩ/K. In the temperature region between 200 and 330 K, the Au film shows the very small MR as shown in **Figure 4**, and therefore, the linear temperature dependence of resistance can be seen in the whole field range up to 56 T. To convert the resistance to temperature, the MRs are fitted by the polynomial function of the temperature (T) and the magnetic field (H):

$$R(T, H) = a_0 + a_1 T + a_2 H + a_3 T^2 + a_4 TH + a_5 H^2 + an, \tag{7}$$

where a_n for n = 0,1,2,... is the polynomial coefficient. The result of the fitting by this polynomial function for n = 0,1,2,...,35 is in good agreement with the data as shown by the rainbow surface in **Figure 4**. The deviation of the fitting function from the data points is within 4 mΩ over the entire range of temperatures and magnetic fields, which corresponds to the error of the temperature of 0.45 K. Hence, this Au film thermometer enables the accurate MCE measurements in the pulsed fields up to 56 T. The validity of this technique was demonstrated in the wide range of temperature through the measurements on Gd at around room temperature and on $Gd_3Ga_5O_{12}$ at low temperatures [36].

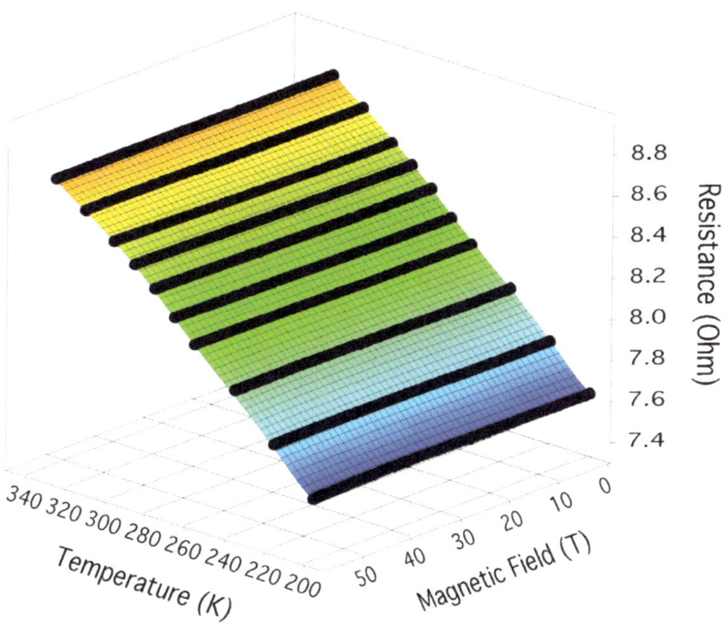

Figure 4. Magnetoresistance of the Au film thermometer deposited on a sapphire disk measured in the pulsed field up to 56 T at the various temperatures (black curves). The rainbow surface is the fitting to the polynomial function discussed in the text.

4. Magnetocaloric properties of metamagnetic shape memory alloys

4.1. Experimental procedures

In this chapter, polycrystalline samples of $Ni_{45}Co_5Mn_{50-x}In_x$ (x = 13.3 and 13.5) and $Ni_{41}Co_9Mn_{31.5}Ga_{18.5}$ were used for the MCE, magnetization and specific heat measurements. The samples of $Ni_{45}Co_5Mn_{50-x}In_x$ (x = 13.3 and 13.5) were prepared by the induction melting method. The ingots were annealed at 900°C for 24 h under argon atmosphere. The samples of $Ni_{41}Co_9Mn_{31.5}Ga_{18.5}$ were prepared by the arc melting method. The ingots vacuum encapsulated in a quartz tube were annealed at 800°C for 72 h and then quenched in cold water.

Magnetization and MCE measurements were performed in pulsed-high-magnetic fields. The pulsed fields are generated by the 56-T magnet at the ISSP. The heat capacity measurements in steady fields were carried out using a thermal-relaxation method in a physical property measurement system, QuantumDesign (PPMS).

4.2. NiCoMnIn

4.2.1. Magnetization measurements

As described in Section 1.2, the metamagnetic shape memory alloy $Ni_{45}Co_5Mn_{36.7}In_{13.3}$ undergoes the MT at around room temperature. **Figures 5(a)** and **(b)** show the isothermal magnetizations of the $Ni_{45}Co_5Mn_{36.7}In_{13.3}$ measured at 310 and 280 K, respectively. The sample is in the ferromagnetic A phase at 310 K, where the M-H curve shows the ferromagnetic behaviour as shown in **Figure 5(a)**. On the other hand, as shown in **Figure 5(b)**, the magnetization increases steeply at the MFIMT, when the magnetic field is applied to the paramagnetic M phase. Here, the transition fields are defined as H_{af} for the field increasing process and H_{ms} for the field decreasing process in **Figure 5(b)**, respectively. The M-H curves measured at the various temperatures are shown in **Figure 6(a)**. One may note that the application of the pulsed-high-magnetic field up to 25 T can complete the MFIMTs in the entire temperature range. From the M-H curve at 4.2 K, the saturation magnetization moment (M_s) in the ferromagnetic A phase is estimated as $M_s = 1.68\ \mu_B$/f.u. Here, μ_B is the Bohr magneton. A formula unit (f.u.) is defined as $Ni_{0.45}Co_{0.05}Mn_{0.367}In_{0.133}$. The magnetic phase diagram of $Ni_{45}Co_5Mn_{36.7}In_{13.3}$ determined from

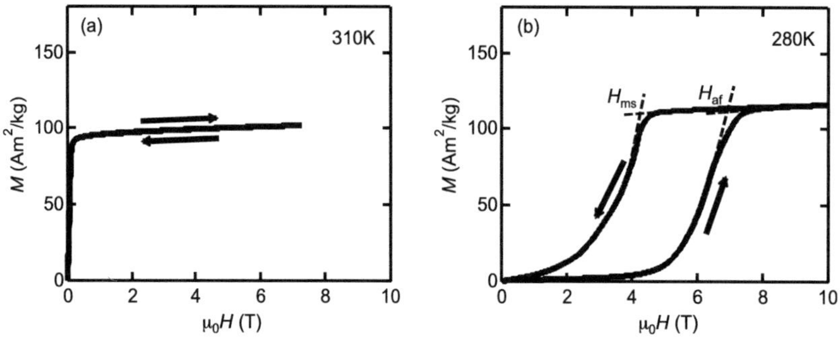

Figure 5. $M-H$ curves measured in the pulsed magnetic fields at (a) 310 and (b) 280 K for $Ni_{45}Co_5Mn_{36.7}In_{13.3}$ [39].

the $M - H$ curves is shown in **Figure 6(b)**. The $M - H$ curves and the magnetic phase diagram for $Ni_{45}Co_5Mn_{36.5}In_{13.5}$ are also provided in **Figures 7(a)** and **(b)**. Hereafter, we denote In13.3 for $Ni_{45}Co_5Mn_{36.7}In_{13.3}$ and In13.5 for $Ni_{45}Co_5Mn_{36.5}In_{13.5}$, respectively. The saturated magnetization in the ferromagnetic A phase for In13.5 is also estimated as M_s = 1.76 μ_B/f.u. As shown in **Figures 6(b)** and **7(b)** and described in Section 1.2, the magnetic phase diagram of $Ni_{45}Co_5 Mn_{50-x}In_x$ is sensitive to the compositional ratio x. However, the saturated magnetizations of the two compositions are similar.

The total entropy change through MFIMT (ΔS_{tot}) can be estimated from the well-known Clausius-Clapeyron equation:

$$\mu_0 \frac{dH_0}{dT} = -\frac{\Delta S_{tot}}{\Delta M}. \tag{8}$$

Here, $\mu_0 = 4\,\pi \times 10^{-7}$ H/m is the space permeability. The ΔM is the difference of magnetization between the A and M phase. The $H_0 = (H_{ms} + H_{af}) / 2$ is used for the transition field. The results are shown in **Figure 8**, where the ΔS_{tot} is defined as the entropy change from the M to the A phase. The ΔS_{tot} at the MT temperature at the zero field is estimated to be 26 J/kg K for In13.3 and 23 J/kg K for In13.5. These values are in good agreement with that obtained from the

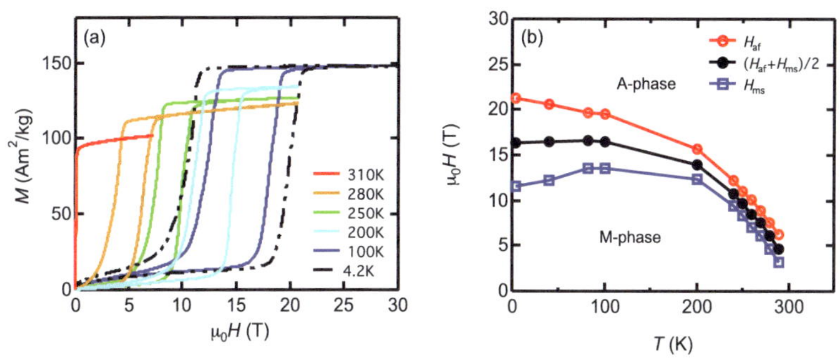

Figure 6. (a) $M - H$ curves measured in the pulsed magnetic fields at the various temperatures and (b) the magnetic phase diagram for $Ni_{45}Co_5Mn_{36.7}In_{13.3}$ (In13.3) [39].

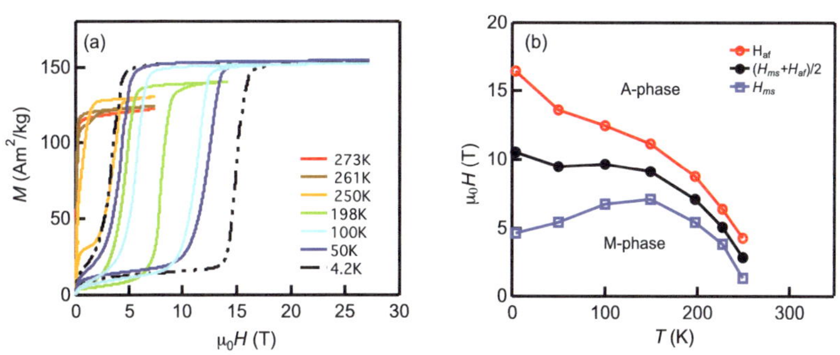

Figure 7. (a) $M - H$ curves measured in the pulsed magnetic fields at the various temperatures and (b) the magnetic phase diagram for $Ni_{45}Co_5Mn_{36.5}In_{13.5}$ (In13.5) [39].

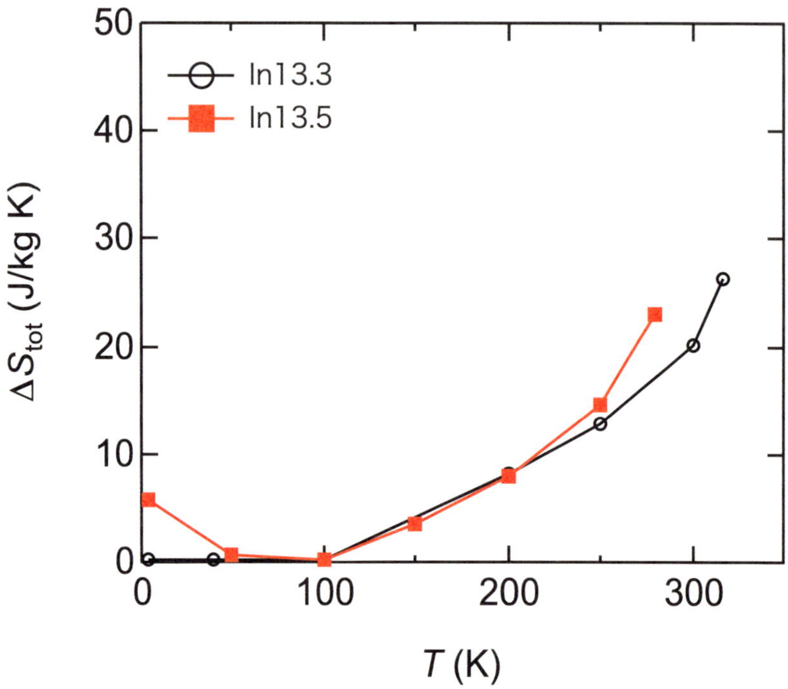

Figure 8. Temperature dependencies of the entropy change at the MFIMT estimated from the magnetic phase diagrams for $Ni_{45}Co_5Mn_{36.7}In_{13.3}$ (In13.3) and $Ni_{45}Co_5Mn_{36.5}In_{13.5}$ (In13.5) [39].

specific heat measurements [39]. The ΔS_{tot} decreases with decreasing temperature in both In13.3 and In13.5 and reaches zero at around 100 K. This indicates that the entropies of A and M phase coincide with each other below 100 K. In other words, the driving force of the MT decreases and reaches to zero below 100 K, which increases the hysteresis of the $M - H$ curves at low temperatures. The MCEs can be calculated by the $M - H$ curves using Eq. (5), which are compared with the results of direct measurements in the next section.

4.2.2. MCE measurements in the pulsed magnetic fields

The direct MCE measurement technique described in Section 3 was first applied to the In13.3 [39]. **Figures 9(a)** and **(b)** show the adiabatic changes of the sample temperature of In13.3 measured at 310 and 280 K. The typical mass of the sample for the MCE measurements is about 50 mg, which is more than 100 times larger than that of the thermometer. Therefore, we can neglect the heat capacity of the thermometer in the measurements. Since the sample is in the ferromagnetic A phase at 310 K, the conventional MCE (monotonic heating/cooling with increasing/decreasing magnetic field) due to the forced spin alignment by the applied field is observed. It is important to note that the reversible temperature change demonstrates the experimental validity: negligibly small heat exchange between the sample and the surroundings and fast response of the thermometer [36]. The result at 280 K is the MCE when the magnetic field is applied to the M phase [**Figure 9(b)**]. The steep cooling (IMCE) occurs at the MFIMT in the field-increasing process. In the successive field-decreasing process, the sample temperature heats up at MFIMT and approximately reaches to the initial temperature at the zero field as

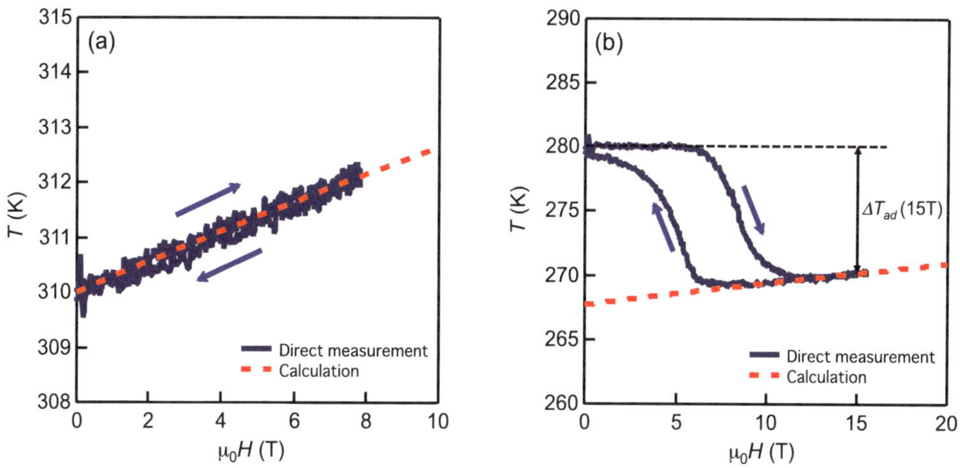

Figure 9. Magnetic field variations of the sample temperature in the adiabatic condition for In13.3. (a) The magnetic field applied to the ferromagnetic A phase and (b) the magnetic field applied to the paramagnetic M phase [39].

expected for the adiabatic measurements. The difference between the initial and final temperatures is less than 0.5 K, which indicates that the heating caused by the hysteresis loss is negligibly smaller than the MCE. In addition, a gentle slope is observed in the field-induced A phase, which is due to the change in the entropy of the ferromagnetic spins. This conventional MCE can be calculated by using the following mean-field model:

$$M(T,H) = Ng\mu_B J B_J\left(\frac{g\mu_B J\mu_0(H-\lambda M)}{k_B T}\right). \tag{9}$$

Here, $N = 6.02 \times 10^{23}$ mol^{-1} is Avogadro's number, g is the g factor, J is the spin quantum number, $B_J(x)$ is the Brillouin function, and $k_B = 1.38 \times 10^{-23}$ J/K is Boltzmann constant. The coefficient λ represents the magnetic interaction, which is defined as $\lambda = 3k_B T_\Theta / Ng^2\mu_B^2 J(J+1)$ by the Weiss temperature T_Θ. Using the experimentally obtained values $g^J = 1.68$ and $T_\Theta = T_C = 387$ K, the magnetization of A phase can be calculated as a function of the temperature and magnetic field. Hence, the MCEs of A phase are calculated by this $M(T, H)$ curve through Eq. (6). The results are plotted as the red-dashed lines in **Figures 9(a)** and **(b)** and are in reasonable agreement with the experimental data. On the other hand, the MCE in the paramagnetic M phase is very small and therefore, cannot be calculated by this model.

Figure 10 shows the temperature dependence of ΔT_{ad} (15T), which is indicated in **Figure 9(b)**, the difference of the sample temperatures between 15 and 0 T as a function of the initial temperature. The ΔT_{ad} (15T) decreases linearly with increasing temperature and is in good agreement with the indirectly estimated values from the magnetization as shown in **Figure 10**.

4.2.3. Specific heat measurements

From the magnetization and MCE measurements, the total entropy change at MFIMT and the magnetic entropy in the field-induced A phase is estimated. In this section, the specific heat

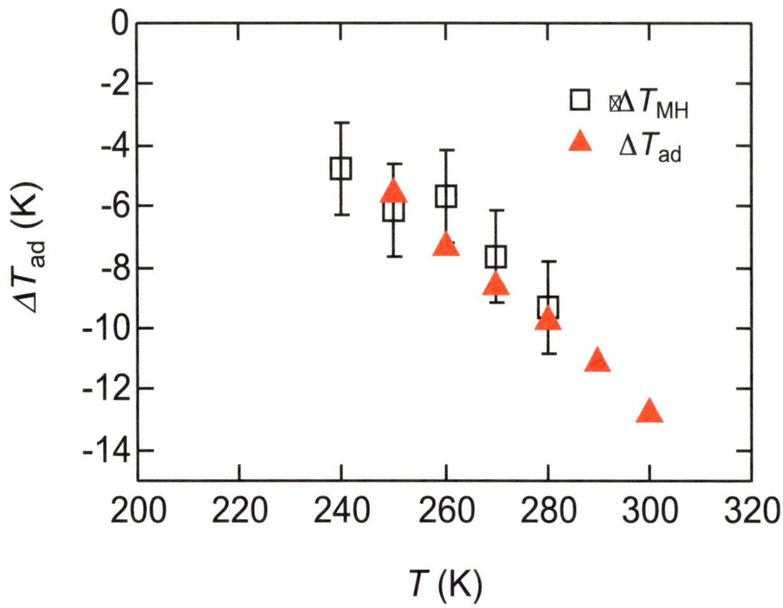

Figure 10. The temperature dependence of ΔT_{ad}(15T) (red triangles), which is defined in Figure 9(b). The open squares are the MCEs calculated from isothermal magnetization curves using Eq. (5) [39].

measurements are performed to evaluate the electronic and lattice contributions to the total entropy. As shown in **Figures 6(b)** and **7(b)**, the samples remain in the A phase at the low temperatures when they are cooled down in the magnetic field above H_{ms}. Therefore, the application of high magnetic fields enables us to measure the specific heat in both A and M phase for a sample. **Figure 11** shows the results of specific heat measurements for In13.5 at low temperatures, which are plotted as C/T versus T^2. The data of A phase (open circles) are measured after the field cooling at 12 T from room temperature. The data in both phases can be fitted by the function:

$$C = \gamma T + \beta T^3, \tag{10}$$

where the first term represents the electronic contribution and the second term corresponds to the phonon contribution. The electronic contribution to the specific heat is characterized by Sommerfeld coefficient, γ, which is proportional to the density of states at Fermi level. The phonon contribution with the coefficient β provides us with the Debye temperature Θ_D. The values [γ, Θ_D] are estimated to be [γ^A = 52 mJ/kg K^2 (3.4 mJ/mol K^2), $\Theta_D^A = 314$ K] for A phase and [γ^M = 48 mJ/kg K^2 (3.1 mJ/mol K^2), $\Theta_D^M = 361$ K] for M phase, respectively. The electronic contribution to the total entropy change ΔS_{ele} is estimated as:

$$\Delta S_{ele}(T) = (\gamma^A - \gamma^M)T. \tag{11}$$

At 300 K, ΔS_{ele} becomes 1.2 J/kg K. This value corresponds to about 5% of the total entropy change at MT. Thus, the electronic contribution to the IMCE does not play dominant role.

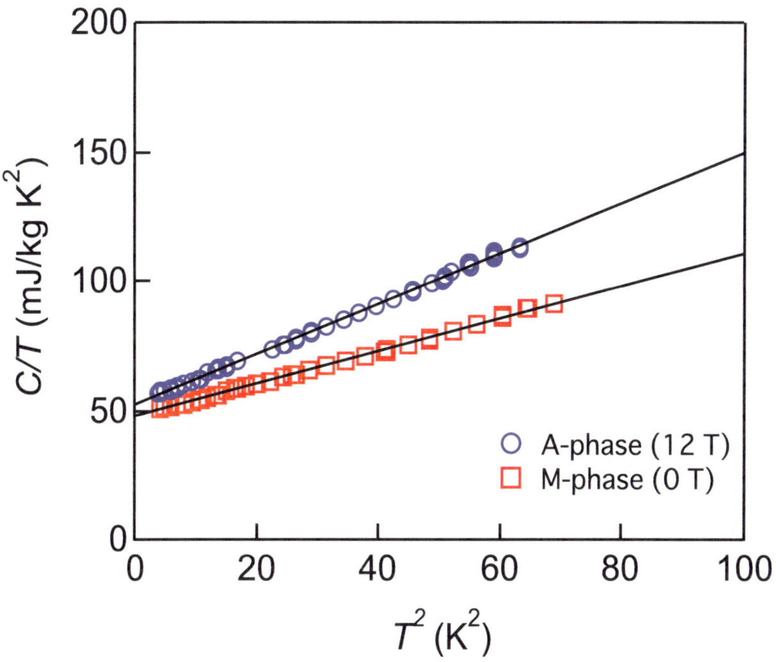

Figure 11. Specific heat of In13.5 plotted as C/T versus T^2 [39].

The specific heat of lattice system can be calculated by assuming the Debye model:

$$C_V = 9Nk_B\left(\frac{T}{\Theta_D}\right)^3\int_0^{\Theta_D/T}\frac{x^4e^x}{(e^x-1)^2}dx. \tag{12}$$

Therefore, the lattice contribution to the total entropy change $\Delta S_{lat}(T)$ due to the difference of the Debye temperature between Λ and M phase can be calculated. Here, the C_V in Eq. (10) is the specific heat under the constant volume, while the measurements are carried out in the isobaric condition. Hence, one has to consider the difference of the specific heat between at the constant pressure and at the constant volume conditions, which can be calculated as:

$$C_p - C_V = 9\alpha^2BVT, \tag{13}$$

where α is bulk thermal expansion coefficient, B is bulk elastic modulus, and V is molar volume. Using the values α = 6.67 × 10^{-6} K^{-1} for $Ni_{45}Co_5Mn_{37}In_{13}$ [40] and B =140 GPa for $Ni_{50}Mn_{25}In_{25}$ [41], one obtains $C_p - C_V$ = 2.3 J/kg K at 300 K, which is smaller than the 0.6% of the experimentally obtained C_p at 300 K [39]. Therefore, the calculated data C_{lat} can be compared with the C_p for the entire temperature range. The ΔS_{lat} is calculated as 51 J/kg K at 300 K, which is about twice larger than ΔS_{tot} value obtained from the phase diagram [**Figure 7(b)**]. This indicates that the lattice contribution plays the central role in the IMCE.

4.2.4. Estimations of the electronic, lattice and magnetic entropy changes

Using Eqs. (11) and (12), ΔS_{ele} and ΔS_{lat} can be calculated for the entire temperature range, which are plotted as the blue- and green-solid curves in **Figure 12**. The magnetic entropy change ΔS_{mag} is calculated by subtracting ΔS_{ele} and ΔS_{lat} from the total entropy change ($\Delta S_{mag} = \Delta S_{tot} - \Delta S_{ele} - \Delta S_{lat}$). The T_{MT} dependence of ΔS_{mag} estimated from the experimental data of ΔS_{tot} below 300 K is plotted as the red-dashed curve in **Figure 12**. The ΔS_{mag} is estimated as $\Delta S_{mag} = -29$ J/kg K (1.9 J/mol K) at 300 K. As mentioned in Section 4.2.2, the magnetic entropy in A phase (S^{A}_{mag}) can be calculated by using the mean-field model. Using the values $T_C = 383$ K and $J = 0.88$ for In13.5, the S^{A}_{mag} is calculated as 79 J/kg K (5.1 J/mol K) at 300 K. Therefore, the magnetic entropy of the M phase can be estimated as $S^{M}_{mag} = 108$ J/kg K (7.0 J/mol K). If one assumes a random arrangement of spins in the paramagnetic M phase, the magnetic moment of the M phase $gJ_M = 1.32$ is obtained from the relation: $S^{M}_{mag} = Nk_B \ln(2J_M + 1)$. This magnetic moment is smaller than that of the A phase by 25%. This result indicates that the magnetic contribution to the total entropy change is composed of not only the negative change due to the spin ordering but also the positive change due to the change in the magnetic moment.

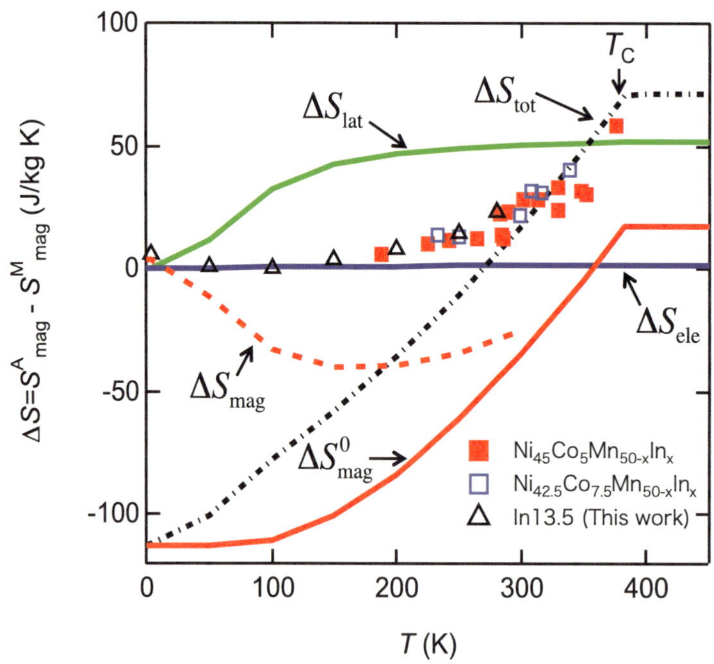

Figure 12. Calculated entropy changes through the MT as functions of the transition temperature (ΔS_{ele}, ΔS_{lat}, ΔS_{mag}, and ΔS_{tot} versus T_{MT}) [39]. The solid and open squares are the experimentally obtained ΔS_{tot} [22, 42]. The open triangles are the ΔS_{tot} estimated by the $M - H$ curves shown in Figure 7(a).

By assuming the constant value of $S^{M}_{mag} = 108$ J/kg K, the temperature dependence of the magnetic entropy change can be calculated as $\Delta S^{0}_{mag}(T) = S^{A}_{mag}(T) - S^{M}_{mag}$. The result is plotted as a red-solid curve in **Figure 12**. The calculated ΔS_{tot} is also plotted as the black-dashed curve in **Figure 12**, which coincides with the experimentally obtained data for $Ni_{50-y}Co_yMn_{50-x}In_x$ for the various x and y values above 300 K. Above T_C, the ΔS_{tot} shows the small temperature dependence. In this temperature region, both A and M phase are paramagnetic, where the contribution of the spin ordering can be neglected. Hence, the ΔS_{tot} was determined by the difference of the lattice entropy and the difference of the magnetic moment. Below T_C, the A phase is ferromagnetic, and therefore, the contribution of the spin alignment increases with decreasing T_{MT}. This effect dominates the T_{MT} dependence of the ΔS_{tot}. This behaviour of the ΔS_{tot} in the vicinity of the T_C can be confirmed in the x dependencies of the $Ni_{50-y}Co_yMn_{50-x}In_x$ [22, 43].

Below 300 K, the calculated ΔS_{tot} no longer reproduces the experimental results. This discrepancy can be attributed to the improper assumption of the constant S^{M}_{mag}. To obtain the deeper insight into the magnetic contribution to the IMCE at lower temperatures, the detailed information about the magnetic structure of M phase is necessary.

4.3. NiCoMnGa

In the NiCoMnIn, the negative entropy change due to the spin ordering through the MFIMT suppresses the ΔS_{tot} and dominates the T_{MT} dependence of the ΔS_{tot}, as shown in **Figure 12**. Therefore, the large IMCE due to the lattice entropy change and due to the change of magnetic moment is expected to occur when a material undergoes MT from the ferromagnetic/paramagnetic M phase to the ferromagnetic/paramagnetic A phase.

Recently, Kihara et al. carried out the magnetization and MCE measurements for $Ni_{41}Co_9Mn_{31.5}Ga_{18.5}$. **Figure 13** shows the isothermal $M - H$ curve measured in the pulsed magnetic field at 260 K. At this temperature, the sample is in ferromagnetic M phase at zero-field. As shown in **Figure 13**, the sample undergoes the MFIMT from the ferromagnetic M phase to the ferromagnetic A phase. In this case, it can be predicted that the negative entropy change due to the spin ordering is smaller than that when the magnetic field is applied to the paramagnetic M phase. Hence, the large and T_{MT} independent IMCE is expected to be observed.

Figure 14 shows the result of the direct MCE measurement at 257.5 K. The sample temperature increases with increasing magnetic field below 16 T in both field increasing (the M phase) and decreasing (the field induced A phase) processes. This temperature increase can be considered as the conventional MCEs due to the forced spin alignment by the magnetic field. In the field increasing process, the sample temperature steeply decreases above 16 T, which can be considered as the IMCE through the MFIMT. However, this temperature decrease is observed up to 36 T, where the MFIMT seems to be completed above 30 T according to the magnetization measurement (**Figure 13**). In the successive field decreasing process, the sample temperature increases reversibly and starts decreasing at around 25 T, where there is no anomaly in the $M - H$

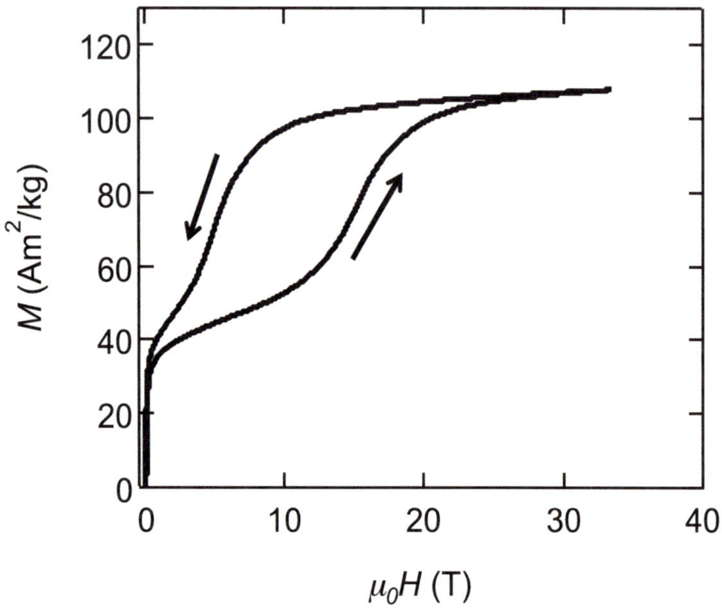

Figure 13. $M - H$ curve measured in the pulsed magnetic field at 260 K for $Ni_{41}Co_9Mn_{31.5}Ga_{18.5}$.

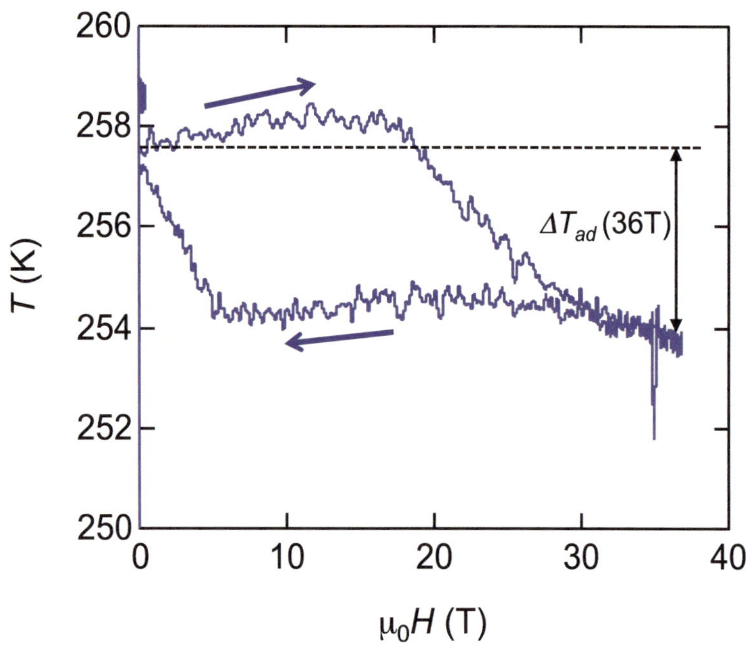

Figure 14. Magnetic field variation of the sample temperature in the adiabatic condition for $Ni_{41}Co_9Mn_{31.5}Ga_{18.5}$ measured at the initial temperature of 257.5 K.

curve. The temperature difference between 0 and 36 T is ΔT_{ad} (36T) = −3.9 K, which is indicated in **Figure 14**. Contrary to the expectation, the large IMCE is not observed. The origin of this unexpected MCE in the $Ni_{41}Co_9Mn_{31.5}Ga_{18.5}$ is still unclear.

5. Conclusion

In this chapter, the direct measurement system of magnetocaloric effect (MCE) under pulsed-high-magnetic field is introduced. Using this technique, the inverse magnetocaloric effects (IMCE) of NiCoMnIn and NiCoMnGa were measured as adiabatic temperature changes in pulsed fields. The electronic, lattice and magnetic contributions to the total entropy change through the magnetic field induced martensitic transformation (MFIMT) are individually evaluated by the combination of the results of MCE, specific heat and magnetization measurements. Through the analysis for NiCoMnIn, the validity of the present experiments for the metamagnetic shape memory alloys is demonstrated.

In conclusion, the origins of the IMCE of NiCoMnIn are as follows. The electronic contribution to the IMCE is negligibly small. On the other hand, the significant change in entropy of the lattice system plays a dominant role in the IMCE. The magnetic contribution involves the two different contributions competing with each other: the increase of magnetic moment (positive entropy change) and the spin ordering (negative entropy change). Moreover, the magnetic contribution determines the composition dependence of the total entropy change (ΔS_{tot}) at the martensitic transformation (MT), when the MT temperature (T_{MT}) and Curie temperature in the austenite phase (T_C) are close to each other. However, the model presented in this chapter cannot reproduce the composition dependence of the ΔS_{tot} when the T_{MT} and T_C are far away from each other. To clarify the origin of the IMCE for the entire temperature range, the detailed experiments about the magnetic structure of the martensitic phase are important.

The magnetization and MCE measurements are carried out at 260 K for $Ni_{41}Co_9Mn_{31.5}Ga_{18.5}$. In the MCE measurement, a complex field dependence of the sample temperature is observed. To clarify the origin of the MCE in the $Ni_{41}Co_9Mn_{31.5}Ga_{18.5}$, the MCE measurements in the wide range of temperatures and the instantaneous observation of the structural change through the MFIMT are highly desirable.

Acknowledgements

The authors thank Dr A. Miyake for helping with the magnetization measurements under the pulsed magnetic fields at the Institute for Solid State Physics, the University of Tokyo. This chapter was partly supported by the Ministry of Education, Culture, Sports, Science and Technology, Japan, through Grant-in-Aid for Scientific Researches (Grant Nos. 23340096 and 16K17734).

Author details

Takumi Kihara[1]*, Xiao Xu[2], Wataru Ito[3], Ryosuke Kainuma[2], Yoshiya Adachi[4], Takeshi Kanomata[5] and Masashi Tokunaga[6]

*Address all correspondence to: t_kihara@imr.tohoku.ac.jp

1 Institute for Materials Research, Tohoku University, Sendai, Japan

2 Department of Materials Science, Tohoku University, Sendai, Japan

3 Materials Science and Engineering, Sendai National College of Technology, Natori, Japan

4 Graduate School of Science and Engineering, Yamagata University, Yonezawa, Japan

5 Research Institute for Engineering and Technology, Tohoku Gakuin University, Tagajo, Japan

6 The Institute for Solid State Physics, The University of Tokyo, Kashiwa, Japan

References

[1] Debye P. Some observations on magnetization at low temperature. Annals of Physics. 1926;**81**:1154

[2] Giauque WF. A thermodynamic treatment of certain magnetic effects. A proposed method of producing temperatures considerably below 1 degree absolute. Journal of the American Chemical Society. 1927;**49**(8):1870

[3] Pecharsky VK, Gschneidner KA Jr. Giant magnetocaloric effect in $Gd_5(Si_2Ge_2)$. Physical Review Letters. 1997;**78**:4494

[4] Brück E. Developments in magnetocaloric refrigeration. Journal of Physics D. 2005;**38**:R381

[5] Gschneidner KA Jr, Pecharsky VK, Tsokol AO. Recent developments in magnetocaloric materials. Reports on Progress in Physics. 2005;**68**:1479

[6] Choe W, Pecharsky VK, Pecharsky AO, Gschneidner KA Jr, Young VG Jr, Miller GJ. Making and breaking covalent bonds across the magnetic transition in the giant magnetocaloric material $Gd_5(Si_2Ge_2)$. Physical Review Letters. 2000;**84**:4617

[7] Morellon L, Arnold Z, Magen C, Ritter C, Prokhnenko O, Skorokhod Y, Algarabel PA, Ibarra MR, Kamarad J. Pressure enhancement of the giant magnetocaloric effect in $Tb_5Si_2Ge_2$. Physical Review Letters. 2004;**93**:137201

[8] Tegus O, Bruck E, Buschow KHJ, de Boer FR. Transition-metal-based magnetic refrigerants for room temperature applications. Nature (London). 2002;**415**:150

[9] Fujieda S, Fujita A, Fukamichi K. Large magnetocaloric effect in $La(Fe_xSi_{1-x})_{13}$ itinerant-electron metamagnetic compounds. Applied Physics Letters. 2002;**81**:1276

[10] Wada H, Tanabe Y. Giant magnetocaloric effect of $MnAs_{1-x}Sb_x$. Applied Physics Letters. 2001;**58**:3657

[11] Fujita A, Fujieda S, Hasegawa Y, Fukamichi K. Itinerant-electron metamagnetic transition and large magnetocaloric effects in $La(Fe_xSi_{1-x})_{13}$ compounds and their hydrides. Physical Review B. 2003;**67**:104416

[12] Pecharsky VK, Gschneidner KA Jr. Magnetocaloric effect and magnetic refrigeration. Journal of Magnetism and Magnetic Materials. 1999;**200**:44

[13] Krenke T, Duman E, Acet M, Wassermann EF, Moya X, Mañosa L, Planes A. Inverse magnetocaloric effect in ferromagnetic Ni-Mn-Sn alloys. Nature Materials. 2005;**4**:450

[14] Kainuma R, Imano Y, Ito W, Sutou Y, Morito H, Okamoto S, Kitakami O, Oikawa K, Fujita A, Kanomata T, Ishida K. Magnetic-field-induced shape recovery by reverse phase transformation. Nature (London). 2006;**439**:957

[15] Planes A, Mañosa L, Acet M. Magnetocaloric effect and its relation to shape-memory properties in ferromagnetic Heusler alloys. Journal of Physics: Condensed Matter. 2009;**21**:233201

[16] Krenke T, Acet M, Wassermann EF. Ferromagnetism in the austenitic and martensitic states of Ni-Mn-In alloy. Physical Review B. 2006;**73**:174413

[17] Liu J, Gottschall T, Skokov KP, Moore JD, Gutfleisch O. Giant magnetocaloric effect driven by structural transitions. Nature Materials. 2012;**11**:620–626

[18] Webster PJ, Ziebeck KRA, Town SL, Peak MS. Magnetic order and phase transition in NiMnGa. Philosophical Magazine Part B. 1984;**49**:295

[19] Krenke T, Acet M, Wassermann EF. Martensitic transitions and the nature of ferromagnetism in the austenitic and martensitic states of Ni-Mn-Sn alloys. Physical Review B. 2005;**72**:014412

[20] Sutou Y, Imano Y, Koeda N, Omori T, Kainuma R, Ishida K. Magnetic and martensitic transformations of NiMnX (X = In, Sn, Sb) ferromagnetic shape memory alloys. Applied Physics Letters. 2004;**85**:4358

[21] Planes A, Mañosa L, Hennry Ehrenreich and Frans Spaepen. Vibrational properties of shape-memory alloys. In: Solid State Physics. Elsevier, Amsterdam, Netherlands; 2001. pp. 159–267

[22] Ito W, Imano Y, Kainuma R, Sutou Y, Oikawa K, Ishida K. Martensitic and magnetic transformation behavious in Heusler-type NiMnIn and NiCoMnIn metamagnetic shape memory alloys. Metallurgical and Materials Transactions A. 2007;**38A**:759

[23] Sakon T, Kitaoka T, Tanaka K, Nakagawa K, Nojiri H, Adachi Y, Kanomata T, Vadim Glebovsky. Magnetocaloric and magnetic properties of meta-magnetic Heusler alloy $Ni_{41}Co_9Mn_{31.5}Ga_{18.5}$. In: Progress in Metallic Alloys. InTech, Rijeka, Croatia; 2016. pp. 265–287

[24] Tishin AM. Magnetocaloric effect in the vicinity of phase transition. In: Buschow KHJ, editor. Handbook of Magnetic Materials. Elsevier, Amsterdam, Netherlands; 1999. pp. 395–524

[25] Pecharsky VK, Gschneidner KA Jr, Pecharsky AO, Tishin AM. Thermodynamics of the magnetocaloric effect. Physical Review B. 2001;**64**:144406-1-13

[26] Amaral JS, Amaral VS. On estimating the magnetocaloric effect from magnetization measurements. Journal of Magnetism and Magnetic Materials. 2010;**322**:1552–1557

[27] Kataoka M, Kanomata T, Umetsu RY, Kainuma R. Normal and inverse magnetocaloric effects in the ferromagnetic shape memory alloys. Journal of Magnetism and Magnetic Materials. 2014;**361**:34–43

[28] Balli M, Fruchart D, Gignoux D, Zach R. The "colossal" magnetocaloric effect in $Mn_{1-x}Fe_xAs$: What are we really measuring. Applied Physics Letters. 2009;**95**:072509-1-3

[29] Cui W, Liu W, Zhang Z. The origin of large overestimation of the magnetic entropy changes calculated directly by Maxwell relation. Applied Physics Letters. 2010;**96**:222509-1-3

[30] Tocado L, Palacios E, Burriel R. Entropy determinations and magnetocaloric parameters in systems with first-order transitions: Study of MnAs. Journal of Applied Physics. 2009;**105**:093918-1-3

[31] Liu GJ, Sun R, Shen J, Gao B, Zhang HW, Hu FX, Shen BG. Determination of the entropy changes in the compounds with a first-order magnetic transition. Applied Physics Letters. 2007;**90**:032507-1-3

[32] Kuz'min MD, Tishin AM. Magnetocaloric effect. Part 1: An introduction to various aspects of theory and practice. Cryogenics. 1992;**32**(6):545–557

[33] Levitin RZ, Snegirev VV, Kopylov AV, Lagutin AS, Gerber A. Magnetic method of magnetocaloric effect determination in high pulsed magnetic fields. Journal of Magnetism and Magnetic Materials. 1997;**170**:223–227

[34] Dan'kov SY, Tishin AM, Pecharsky VK, Gschneidner KA. Experimental device for studying the magnetocaloric effect in pulse magnetic fields. Review of Scientific Instruments. 1997;**68**(6):2432–2437

[35] Kohama Y, Marcenat C, Klein T, Jaime M. AC measurement of heat capacity and magnetocaloric effect for pulsed magnetic fields. Review of Scientific Instruments. 2010;**81**:104902-1-13

[36] Kihara T, Kohama Y, Hashimoto Y, Katsumoto S, Tokunaga M. Adiabatic measurements of magneto-caloric effects in pulsed high magnetic fields up to 55 T. Review of Scientific Instruments. 2013;**84**:074901-1-7

[37] Kohama Y, Hashimoto Y, Katsumoto S, Tokunaga M, Kindo K. Heat-pulse measurements of specific heat in 36 ms pulsed magnetic fields. Measurement Science and Technology. 2013;**24**:115005

[38] Kindo K. 100 T magnet developed in Osaka. Physica B. 2001;**294–295**:585–590

[39] Kihara T, Xu X, Ito W, Kainuma R, Tokunaga M. Direct measurements of inverse magnetocaloric effects in metamagnetic shape-memory alloy NiCoMnIn. Physical Review B. 2014;**90**:214409-1-6

[40] Li Z, Jing C, Zhang HL, Yu DH, Chen L, Kang BJ, Cao SX, Zhang JC. A large and reproducible metamagnetic shape memory effect in polycrystalline $Ni_{45}Co_5Mn_{37}In_{13}$ Heusler alloy. Journal of Applied Physics. 2010;**108**:113908-1-4

[41] Qawasmeh Y, Hamad B. Investigation of the structural, electronic, and magnetic properties of Ni-based Heusler alloys from first principles. Journal of Applied Physics. 2012;**111**: 033905-1-6

[42] Kustov S, Corró ML, Pons J, Cesari E. Entropy change and effect of magnetic field on martensitic transformation in a metamagentic Ni-Co-Mn-In shape memory alloy. Applied Physics Letters. 2009;**94**:191901-1-3

[43] Gottshall T, Skokov KP, Benke D, Gruner ME, Gutfleisch O. Contradictory role of the magnetic contribution in inverse magnetocaloric Heusler materials. Physical Review B. 2016;**93**:184431-1-5

Chapter 4

Applications of Nanostructural NiTi Alloys for Medical Devices

Elena O. Nasakina, Mikhail A. Sevostyanov,
Alexander S. Baikin, Alexey V. Seryogin,
Sergey V. Konushkin, Konstantin V. Sergienko,
Alexander V. Leonov and Alexey G. Kolmakov

Additional information is available at the end of the chapter

http://dx.doi.org/10.5772/intechopen.69238

Abstract

New nanostructural shape memory alloy (55.91 wt% of Ni and 44.03 wt% of Ti) for the production of minimally invasive implantation medical devices (stents) was tested for corrosion resistance under static conditions by dipping it into solutions with various acidities (pH from 1.68 to 9.18) for 2 years, for static mechanical properties and for biocompatibility. The material for investigations was 280-μm wires before and after thermal treatment at 450°C for 15 min in air and surface mechanical treatment. The characteristic image and size of grains were determined using the transmission electron microscope (TEM), and the phase composition; surface morphology; and the layer-by-layer composition were investigated using an X-ray diffractometer; a scanning electron microscope (SEM); and an Auger spectrometer. The nickel release from the investigated nanostructural nitinol is less in comparison with data for microstructural nitinol in a solution of any acidity. Dissolution in the alkali medium is absent. A significant retardation of the nickel ion release (and insignificant concentration as a whole) and the absence of titanium ion release in the weakly acidic and neutral solutions with polished samples are observed. A simultaneous 7–11% increase in strength and plasticity in comparison with microstructural nitinol was attained. Toxicity of samples has not been revealed.

Keywords: nitinol, biocompatibility, corrosion resistance, shape memory effect, superelasticity

1. Introduction

Nitinol (NiTi) alloys possessing a shape memory effect (SME) and mechanical characteristics similar to the behavior of living tissues have been already used for years as the material for production of medical devices, including implants, for example stents [1–6].

SME promotes the production of the least traumatic self-expanding stents, which are implanted into the human body to relieve obstruction or constriction of the respiratory, digestive, excretory, and cardiovascular systems and to restore the normal circulation of physiological flows without the need for additional devices except the catheter carrier [6]. But nitinol contains nickel (including on its surface) which is toxic for organisms [7–9]. Different authors give completely different durations (up to constant) and magnitudes (from fractions to hundreds of mg/L) of the nickel ion release from microstructural nitinol into the medium [8–16]. And also, a different level of nitinol biocompatibility and of its electro-chemical corrosion characteristics is noted [2, 8–10, 17, 18]. Nitinol properties are generally determined by the structure and the composition, which depend on the production process and treatments [2, 8–9, 11, 17, 19–22].

As is well known, formation of nanostructures imparts special, controlled characteristics to materials [23, 24]. Nitinol, as titanium, is a self-passivated material, that is, it forms a complex surface oxide layer, which protects the material from corrosion, and also is biocompatible in itself [2, 10–12]. This oxide layer also surrounds each grain. Therefore, it is possible to assume that increase in a volume fraction of such grain boundaries in alloy at its nanostructuring will positively affect its corrosion resistance and biocompatibility. On the other hand, the high density of intergranular surface defects could lead to a poor corrosion performance since corrosion attack typically initiates at surface heterogeneities [25].

As described in Ref. [25], one of the most efficient methods of fabrication of bulk nanocrystalline materials is the severe plastic deformation (SPD) leading to the breakdown of coarse grains into nanosized (with the size less than 100 nm) grains. The main SPD techniques are equal channel angular extrusion (ECAP) including thermomechanical treatment, high pressure torsion (HPT), hydrostatic extrusion, and others.

It is noticed in literature that nanostructure has an ambiguous effect on the nitinol properties and this is little studied. Investigations showed that in HCl nitinol in the nanosized state (grain diameter of 10 nm) reveals a significantly lower corrosion resistance than in the microstructural state because of an increase in the boundary length and in the amount of defects of the grain structure; and nitinol in NaCl, on the contrary, is more passive (consequently corrosion resistant) [26]. In the study [25, 27–29], there is no difference in corrosion processes between micro- and nanostructured nitinol. But in case of nanostructured titanium, corrosion resistance reduction was observed [30] though in all works positive influence on mechanical properties of materials was noted. In Refs. [31–33], large recoverable strains, resistance against cyclic plasticity mechanisms, plasticity, and good fatigue response were achieved by grain refinement by means of ECAP. HPT produced an ultrafine-grained structure with grain sizes in the range from 5 to 100 nm, which leads to a very high strength, good ductility, high recovery stress, and to a maximum reverse strain, significantly improving cyclic endurance and pseudoelasticity [34]. Additional electropulse current treatment of coarse-grained and ultrafine-grained NiTi (grain diameter more than 100 nm) results in the formation of structures with a grain size less than 80 and 100 nm, correspondingly (was more effective for microstructural material), with increased deformability and mechanical properties [35]. At the same time, according to the computational model and the experimental data in Ref. [36], it was shown that when the grain size is smaller than some critical value (around 50–80 nm in all directions), the martensitic phase transformation is totally suppressed even though the material severely deforms; whereas in another experiment, nanostructure does not prevent phase transformations that are influenced only by aging regime and cooling rate [37]. Authors of Ref. [38] noted that the cold-drawn nitinol contains 50 nm thick grains and possesses quite high tensile strength, hardness, and fatigue life, the maxima of which are reached after heat treatments at 450°C within 15 min; and heat treatments at above 450°C induced recrystallization and grain and precipitate growth. It is shown that the nanostructural morphology of the nitinol and titanium surfaces at the tissue-implant interface positively affects biocompatibility *in vitro* and *in vivo* and enhances functional activity of desirable cell cultures and enzymes without any fibrous tissue intervention. However, not all nanomorphologies on implants imparted a similar biological response [39, 40].

Thus, it is still of interest how nanostructuring can influence operational characteristics of nitinol as a medical (implantation) material. The aim of this chapter is to investigate the composition, structure, and properties of polycrystalline nitinol with nanograins.

2. Obtaining and investigating nanostructural NiTi alloy for medical application

Nanostructured wires with a diameter of 280 μm made of nitinol (50.9 wt% of Ni) were obtained from the Institution of Russian Academy of Sciences, A.A. Baikov Institute of Metallurgy and Material Science, before processing and after mechanical and/or thermal treatments were investigated.

To obtain the material, we used the modernized complex plastic deformation technology developed in the IMET RAS. The corresponding charge was re-melted several times in a vacuum furnace in argon. Ingots were transformed by rolling and rotary forging at a temperature of 750–1000°C into bars with a diameter of up to 4 mm, from which the wire was obtained by step-wise hot drawing through a synthetic diamond die and by intermediate thermal treatment for stabilization of the material structure and removal of mechanical stress.

The mechanical treatment (by abrasive paper from 180 to 1000 grit and finally by GOI (State Optical Institute) paste to a mirror surface) of the wire surface was carried out by an abrasive cloth for the removal of flat indentations and defects in the form of dimples after wire drawing. The decrease in the diameter was to 10 μm in comparison with the original. The thermal treatment of nitinol that allows one to vary static properties and cyclic loadings in operating conditions with a wide range of deformations is extremely important for stabilization of the properties, constraining (shaping) the samples, and successful application of the product. The optimal conditions for the thermal treatment were chosen from the previous researches as

T = 450°C and an annealing time of 15 min (what has coincided with [38]), and carried out in a LOIP LF 7/13_G2 muffler (Russia) equipped with a programmed controller.

To investigate the wire microstructure, the preliminary etching of the surface was carried out in a mixture of 1 mL HF + 2 mL HNO_3 + 47 mL H_2O composition for 2–3 min, upon termination of which the sample was washed several times in distilled water and dried in air. Investigations were carried out on an Axiovert 40 MAT optical metallographic microscope (Carl Zeiss, Germany) with digital image processing. The characteristic image and size of grains were determined with the use of a TECNAI 12 transmission electron microscope (TEM) (FEI COMPANY, USA), and samples were prepared by means of a GATAN 691 ion-etching device (Gatan Inc., USA). To determine the phase composition, we used an Ultima IV X-ray diffractometer (Rigaku Co., Japan) in Cu Kα—radiation on the basis of Bragg-Brentano method. Phase analysis was prepared in the PDXL program complex using the ICDD database. The surface morphology and the layer-by-layer composition were investigated on a scanning election microscope (SEM) VEGA II SBU with the module INCA Energy for energy-dispersive analysis (TESCAN, Czech Republic) and on a JAMP-9500F Auger spectrometer (JEOL Co., Japan) in combination with ion etching at argon bombardment under an angle of 30°.

The bright-field TEM image in **Figure 1a** shows that nitinol grains resemble nanofibers with a cross-section size from 30 to 70 nm, and the longitudinal section size of fibers is several microns. Thus, grains are extended along the wire axis. Rings are seen in the microdiffraction pattern (**Figure 1b**), which point to polycrystalline structure of the sample. The calculation and analysis of the interplanar spacing distances from the electron diffraction patterns show that they correspond to the B2 phase.

Also, according to the energy-dispersive and X-ray structural analyses (**Figure 2a**), the material volume is represented by the base of the B2 phase of TiNi and inclusions of Ti_2Ni

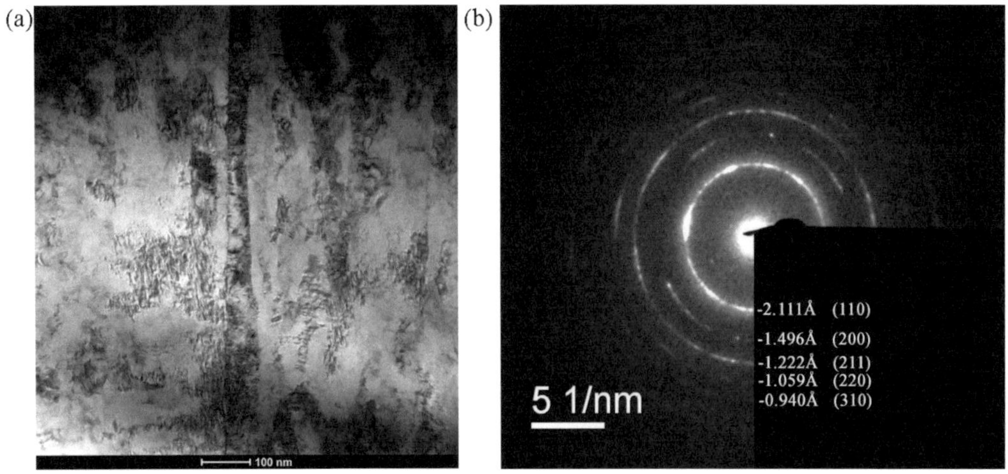

Figure 1. Transmission electron microscopy data: (a) bright-field TEM image and (b) microdiffraction pattern.

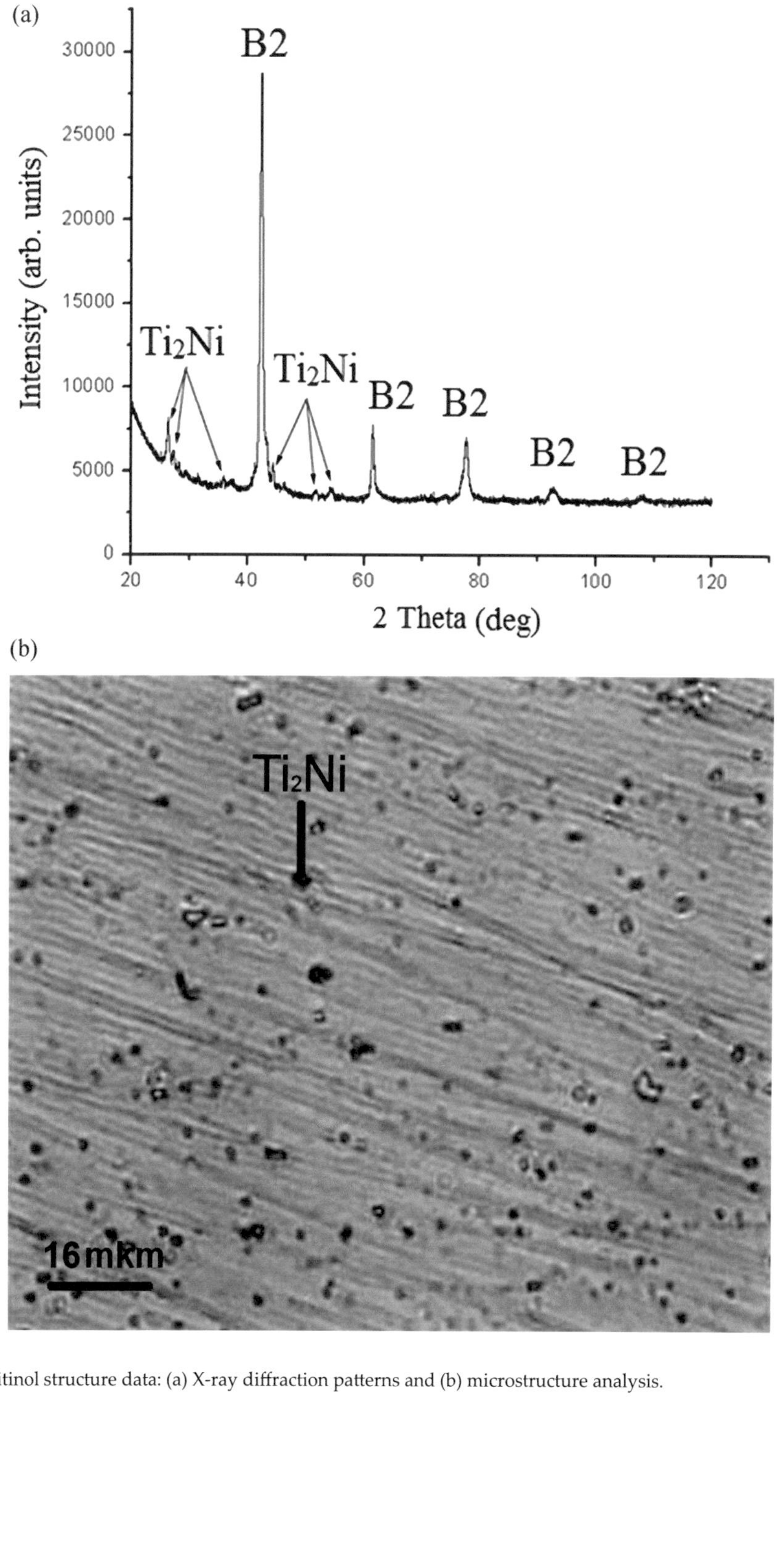

Figure 2. Nitinol structure data: (a) X-ray diffraction patterns and (b) microstructure analysis.

intermetallides. The character of X-ray patterns indicates that the composition is unchanged after thermal treatment. The microstructure analysis (**Figure 2b**) verifies that the base of the alloy investigated is represented by the B2 phase, and the material contains Ti_2Ni intermetallides, and the sizes and the volume fraction of which are unchanged after annealing.

According to the SEM images (**Figure 3a**), the wire surface before treatment is heterogeneous and covered by spots, dark areas alternating with bright ones and the high roughness is clearly expressed. After annealing (**Figure 3b**), the surface externally is similar to the initial one, whereas after polishing (**Figure 3c**) practically all defects and the roughness are smoothed, spots are absent, and only traces of treatment are visible. Surface microdefects are grooves with a depth and width of less than 1 μm, and after polishing and annealing, the wire surface is the most-smoothed and uniform (**Figure 3d**).

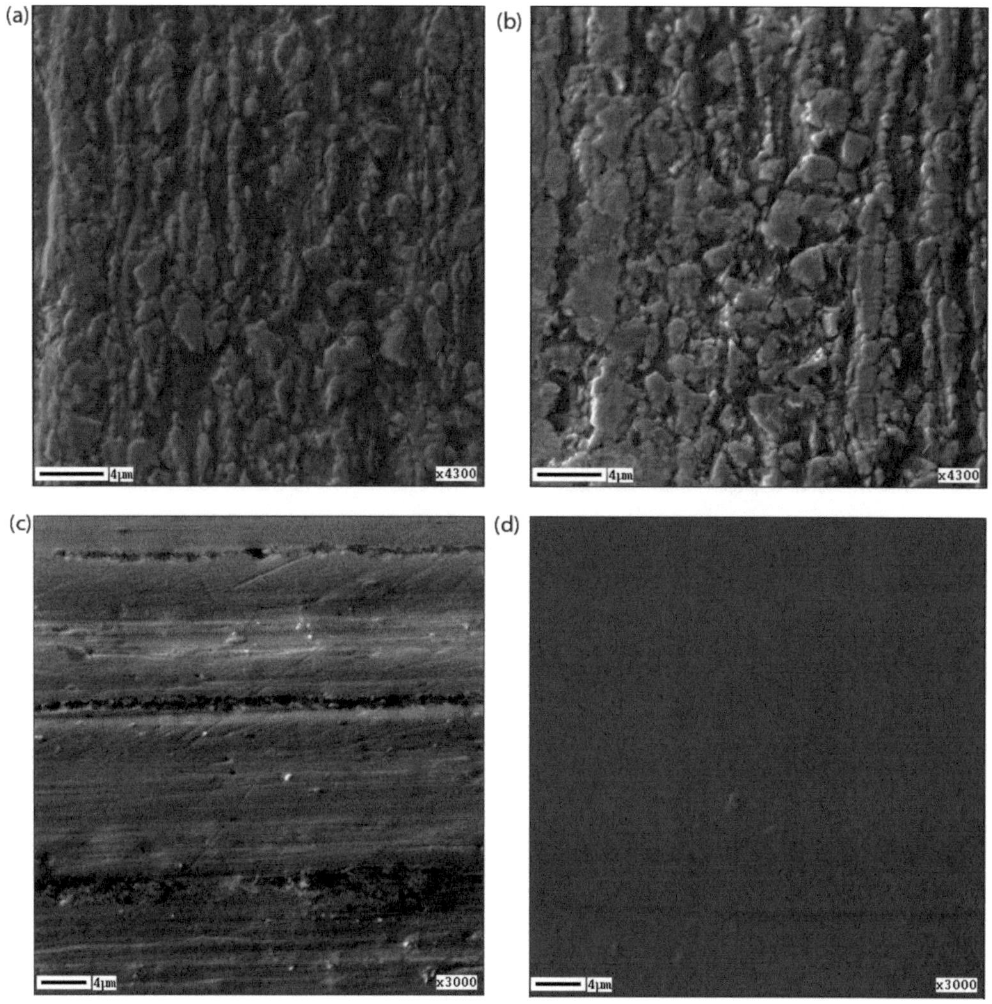

Figure 3. SEM image of the wire surface: (a) before treatment, (b) after annealing, (c) after polishing, and (d) after polishing and annealing.

The compositions of bright and dark spots are different (**Figure 4a**): a high content of titanium oxide is revealed in bright ones and carbon, in dark ones. Both layers attain 3 μm in thickness and are not placed over each other. Such a thick surface layer, as was believed, was a result of the long intermediate thermal treatment during the wire production [8, 9, 41]. Most likely, therefore, annealing for 15 min does not have an effect on features of changing the composition. Carbon is present on the wire surface, probably, due to graphite-containing lubricant (used during wire drawing), which remains on the wire surface and then sticks during annealing. The same effect (impurities on the surface after several cycles of treatment due to lubricant) was observed in Ref. [11]. The composition of the polished surface is homogeneous (**Figure 4b**): the entire wire is covered by an oxide layer less than 20 nm in thickness, which, according to the literature [41], must positively affect corrosion resistance under dynamic conditions because thin oxide layers show greater flexibility and may adapt to loads applied to the material. The minus is the presence of nickel in the surface layer, though in an insignificant amount. The subsequent annealing promotes the formation of an oxide-nitride layer up to 80–150 nm in depth, to 50–60 nm free from nickel that could affect corrosion resistance of material positively (**Figure 4c**), that is also connected with thermal treatment which leads to the formation of mixed surface layer consisting of titanium and nickel oxides and to the growth of the oxide layer thickness [8, 9, 41].

The mechanical properties of samples with a working part length of 45 mm were determined under the conditions of static stretching on an Instron 3382 (Instron, USA) universal testing machine with a loading speed of 2 mm/min. The base diameter was used in the calculation of strength properties. Three to five samples were tested per one experimental point. The conventional yield strength $\sigma_{0.2}$, the ultimate strength σ_u, and the relative elongation δ were determined (**Table 1**).

Results of the mechanical stretching tests showed positive influence of annealing on strength properties of alloy in both structural states, and strength properties of a nanostructural nitinol exceeded strength properties of microstructural nitinol by 1.3–1.5 times. Nitinol wire polishing after drawing promotes an increase in the static properties of a nanostructural alloy of 18%. Relative elongations of all samples were of 51–53%.

Heat treatment significantly increases cyclic durability of nanostructural material (**Figure 5**). Its durability in the initial state is insignificant, and all fatigue curves lie in the low-cyclic area (below $N = 6 \times 10^3$ cycles). After annealing, durability sharply increases both in the field of low-cycle and multicycle fatigue. The durability limit after annealing on the basis of 4×10^6 cycles is 400 MPa.

The wire disintegrates with the formation of a neck (**Figure 6**). The fracture surface is aligned almost perpendicularly to the stretching axis. It consists of a set of unequally sized self-similar pores (cups) of the viscous break. The destruction starts from the most serious defects on the sample surface in the zone of the neck where the main crosscut crack, to which the formed pores are merged, is formed. Heat treatment practically does not influence the nature of destruction.

Micro-Vickers hardness measurements determined at loading 1–2 N by the WOLPERT GROUP 401/402 device—MVD (WILSON Instruments, USA) equipped with a light microscope.

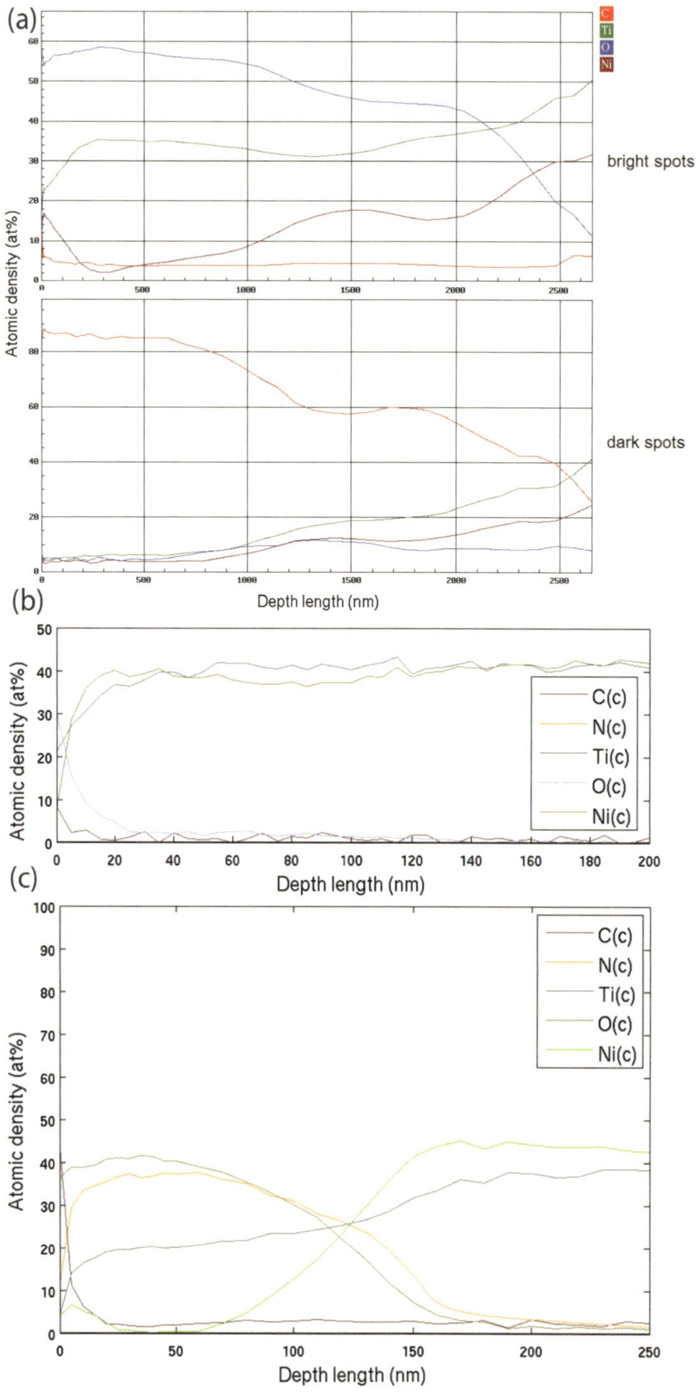

Figure 4. Wire surface structure according to Auger spectroscopy prior to the immersion testing: (a) before treatment and after annealing (dark and bright spots in **Figure 3a** and **b**), (b) after polishing, and (c) after polishing and annealing.

http://dx.doi.org/10.5772/intechopen.69238

Batch number	Sample	$\sigma_{0.2}$, MPa	σ_u, MPa
1	Microstructural nitinol before treatment	497	1423
2	Nanostructural nitinol before treatment	507	1485
3	Microstructural nitinol, annealing of 450°C, 15 min	564	1635
4	Nanostructural nitinol, annealing of 450°C, 15 min	742	1885

Table 1. Effect of annealing on nitinol mechanical properties.

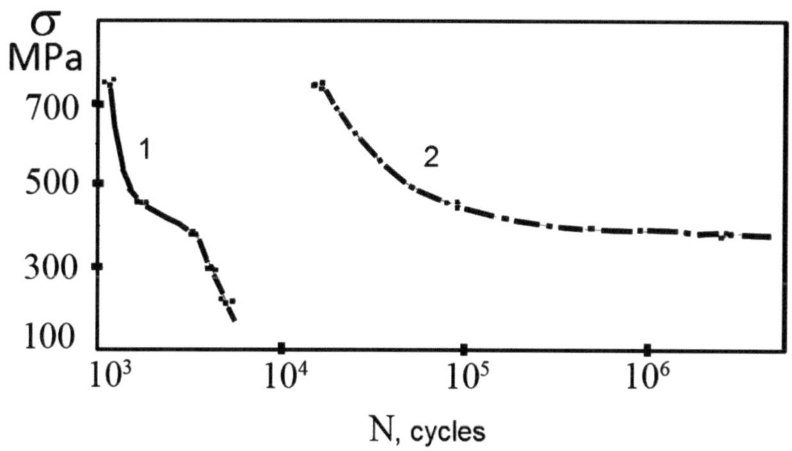

Figure 5. Fatigue curves of a nanostructural nitinol before treatment (1) and after annealing at 450°C, 15 min (2).

Microhardness of nanostructural material is higher by ≈ 38% in comparison with a microstructural analog (332 ± 3 and 240 ± 3 HV, respectively). It can be explained by the increase in volume fraction of grain boundaries and also by the presence of titanium oxides on these borders, the hardness of which is higher than the basic hardness. As a result of surface polishing, a formation of a thin uniform oxide layer with a small share of impurity takes place, and microhardness increases by 17.5% (from 332 ± 3 to 390 ± 4 HV). The subsequent annealing within 15 min at 450°C promotes the thickening of the oxide layer and an increase in the share of titanium oxides in it that leads to additional increase in microhardness by 2.3% (to 399 ± 3 HV).

The corrosion resistance of the material was determined under static conditions by immersion into solutions with various acidities because pH in the human body changes from 1 to 9 (for example, 1.05 at the duodenum; 1.53–1.67 is the norm of gastric juice; 3.8–4 in the near-surface area of the bowels; 7.34–7.43 is the norm of blood; 8.5–9 in the bowels; etc.). We used a neutral 0.9 wt% solution of sodium chloride, artificial plasma, and four standard buffer solutions (to reproduce acidic and alkaline media at the given level) [8, 9, 42], which are listed in **Table 2**, and for comparison, we also used a 0.03-M solution of hydrochloric acid. The standard buffer solutions were prepared from corresponding standard trimetric substances (fixanals) made by Merk (USA).

Wire samples (1—TiNi before treatment, 2—TiNi after annealing, 3—TiNi after polishing, and 4—TiNi after polishing and annealing), with a weight of 32.6 mg each (separately from

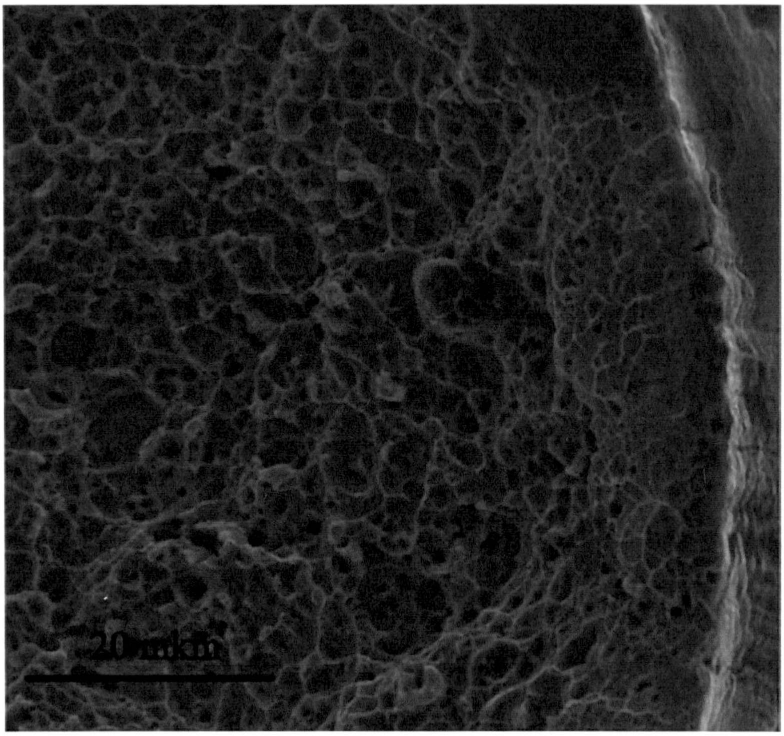

Figure 6. Disintegration behavior of the nitinol wire at static stretching.

No	pH	Composition
1	1.68	Potassium tetraoxalate $KH_3C_4O_8 \times 2H_2O$, 0.05 M
2	3.56	Acid potassium tartrate $C_4H_5O_6K$, 0.025 M
3	4.01	Acid potassium phthalate $C_8H_5O_4K$, 0.05 M
4	6.31	Sodium chloride NaCl, 0.9 wt%
5	9.18	Acid sodium tetraborate $Na_2B_4O_7 \times 10H_2O$, 0.05 M
6	7.36	Artificial plasma: NaCl (92.3 mM), $NaHCO_3$ (26.3 mM), K_2HPO_4 (0.9 mM), KCl (2.7 mM), NaH_2PO_4 (0.22 mM), $CaCl_2$ (2.5 mM), $MgSO_4 \cdot 7H_2O$ (0.82 mM), Na_2SO_4 (1.48 mM), D-glucose $C_6H_{12}O_6$ (5.55 mM) [8–9]

Table 2. pH and composition of the solutions used for immersion tests.

each other) were placed into flasks of heat-resistant laboratory glass (acidic and neutral media) or polypropylene (alkali medium) with 100 mL of the selected solution. The flasks were hermetically closed by a lapped cover and aged in a dark place. Sampling from flasks for analysis was done after a selected period (10, 25, 45, 60, 75, 236 or 287, 704, or 754 days). The initial buffer solutions were used as reference solutions. Analysis was carried out by an ULTIMA 2 sequential atomic emission spectrometry (HORIBA Jobin Yvon, Japan) for using atomic emission spectrometry (AES) with inductively coupled plasma (ICP) for direct simultaneous determination of titanium and nickel in solutions.

After immersion, the surface morphology and layer-by-layer composition were also investigated.

The results of measuring the ion release into solutions are represented in **Figures 7** and **8**. Insignificant corrosion is observed in acidic and neutral media, and the nickel concentration is less than the average magnitudes cited in the literature [8, 9, 11–13, 43]; however, the titanium content is revealed in solutions. There are no results about all samples in the alkaline environment and artificial plasma, and also about TiNi-3 and TiNi-4 samples in solutions with acidity 3.56–6.31 since, in these cases, the release of elements was zero or below a limit of detection for the overall time of the investigation.

Comparison of the treatment effect on the corrosion resistance of samples (**Figure 7**) makes it possible to conclude that the samples undergo the most corrosion after annealing, and the mechanical treatment, as was expected, strongly increases the corrosion resistance of nitinol.

According to the literature, the thermal treatment at a temperature from 400 to 1000°C, which is required for stabilization of the mechanical properties, always results in the significant worsening of the corrosion resistance due to the formation of nonuniform surface layers consisting of titanium and nickel oxides [21, 22, 41]. The undesirable effect of the thermal treatment on the corrosion resistance can also be explained by the occurrence of tempering and recrystallization of the outer cold-hardened (strengthening) layer on the surface of untreated nitinol, which forms during its production. At the same time, the surface treatment, which facilitates the formation of the most perfect and homogeneous passive film, increases the corrosion resistance.

And in case of TiNi-4 and TiNi-3 samples on an initial stage of researches regularity has the return nature. It is possible to assume that in the beginning thicker oxide surface layer of a sample after polishing and annealing acts as the better barrier against diffusion of nickel ions in solution;

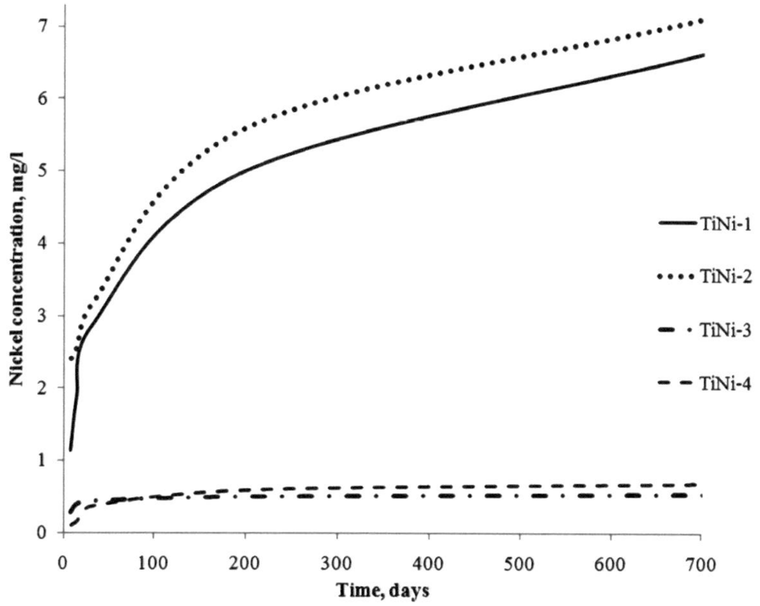

Figure 7. Nickel concentration in solution with pH 1.68 as a function of sample treatment and immersion time.

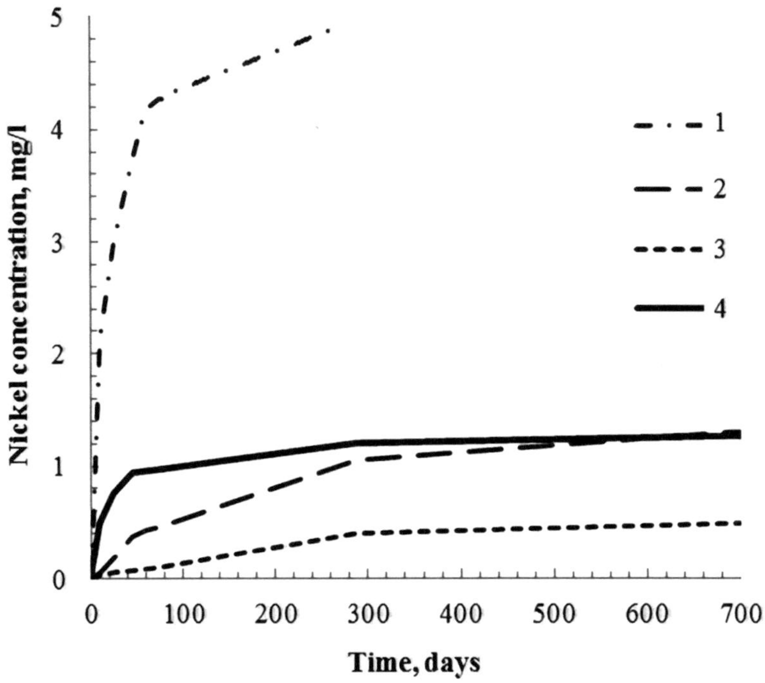

Figure 8. Nickel concentration in investigated solutions with various acidities with nitinol sample without treatment as a function of sampling time. The marked curves correspond to solution pH: 1—1.68, 2—3.56, 3—4.01, and 4—6.31.

however, after a long period of immersion, its nonuniformity does not allow to slow down leaching of elements as effectively as in case of a more uniform passive film which is initially received when polishing. The ratio of metal concentration in solutions with the annealed and the nonannealed samples at existence and lack of polishing treatment approximately coincides.

The surface examination after holding in solutions also reveals the hardest corrosive attack after annealing and the lowest after polishing. However, the wire diameter changes disproportionately to the ion concentration in the solution. For example, the diameters of untreated and annealed wires are almost equal after holding in the media with pH 1.68 (**Table 3**); that is, the surface fails nonuniformly, and it is corroded by single deep holes and pores.

In previous researches [42], a release of ions both of nickel and of titanium was observed; in the most acidic environment, the Ni released is about 1.5–2 times more (however, both concentrations are small, and the titanium concentration is not toxic for an organism), at pH 3.56 nickel concentration was 2–4 times more, and in other solutions titanium was not revealed which is consistent with literary data about its consumption on a passive film formation [8, 9, 42].

The ion concentrations of both nickel and titanium in solutions increase with time; however, this increase has a different character depending on the medium (**Figures 7** and **8**). During short-term tests [42], we believed that the slope of time-dependent concentration curves decreased gradually in all media, flattening out because of the termination of metal etching from the surface. It was so in general, however, the long-term investigation showed small

Diameter, μm	Sample	pH
280.00	Nontreated	–
280.00	Annealed	–
270.00	Polished	–
253.74	Nontreated	1.68
275.59	Nontreated	3.56
277.14	Nontreated	4.01
275.16	Nontreated	6.31/7.36
280.00	Nontreated	9.18
252.46	Annealed	1.68
257.00	Polished	1.68
270.03	Nontreated	1.5 (HCl)

Table 3. Diameters of each type of NiTi wire sample before and after immersion for 2 years in solutions with various pH values.

fluctuations in the slope of the concentration curve toward increase and reduction of metal release in acid media. The last in comparison with published data can be related to sequential processes of the destruction and renewal of the protective oxide film (de- and repassivation) on defect areas [12, 44, 45].

The greater ion release from samples of a single type is observed in the most acidic medium (**Figure 8**), which abruptly drops with increasing pH and again increases in the NaCl solution. This corresponds to expectations. On the one hand, the metal ion release increases with increasing acidity, which in the theory corresponds to an increase in the corrosiveness of the medium. On the other hand, a high metal yield (it is greater than that for solution with acidity of 4.01 and 3.56 in the initial period and for pH 4.01 after 600 days) is revealed in the physiological saline, being a fairly concentrated source of chlorine ions, which are related to corrosion activator ions, having a depassivation and pitting corrosion effect, having a metal affinity greater than oxygen, and displacing the latter from its compounds with metals [46, 47].

Visually, the most damage occurs also at pH 1.68 after holding for 2 years (**Figure 9a** and **Table 3**). According to the location of pitting, the corrosive attack occurs coaxially to the direction of defects in production during drawing. Small traces of corrosion are observed at the surface at an acidity of 3.56. Defects of the metallic surface are no longer revealed at pH 4.01; however, depositions of organic nature are clearly expressed, which cover the wire with great strata and fibers but do not cover it completely (**Figure 9b**). Wires in the neutral solutions are covered by a smooth homogeneous surface layer (**Figure 9c**); and traces of pitting are absent, which supposedly is a result of repassivation of the damaged surface. Changes in the surface and diameter (**Table 3**) are not observed after holding in the alkaline medium. Because the used buffer solutions were not physiological media and served only for formation of the required pH, their effect on the alloy was compared with the effect of hydrochloric acid [42]. The wire surface after long holding in HCl solution (**Figure 9d**) is more like the surface after holding in the physiological saline than in the acidic buffer, and the change in diameter (**Table 3**) is significantly less than that in a solution with pH 1.68. At the same time the concentrations of solved nickel for a short time in solutions of hydrochloric acid with pH 1.56 (the acidity changed to 2 after 2 years of immersion) and the buffer with pH 1.68 were close and that of titanium differed by more than a factor of two (after 10 days nickel concentration in HCl solution was 1.94 mg/l, titanium concentration was 0.515 mg/l). This gives occasion to suppose that the behavior mechanism in the two chloride-containing media is similar, etching of metals from the surface slows down at the initial similarity of magnitudes of solved nickel at long-term holding in the acid, and the surface is covered by a protective film.

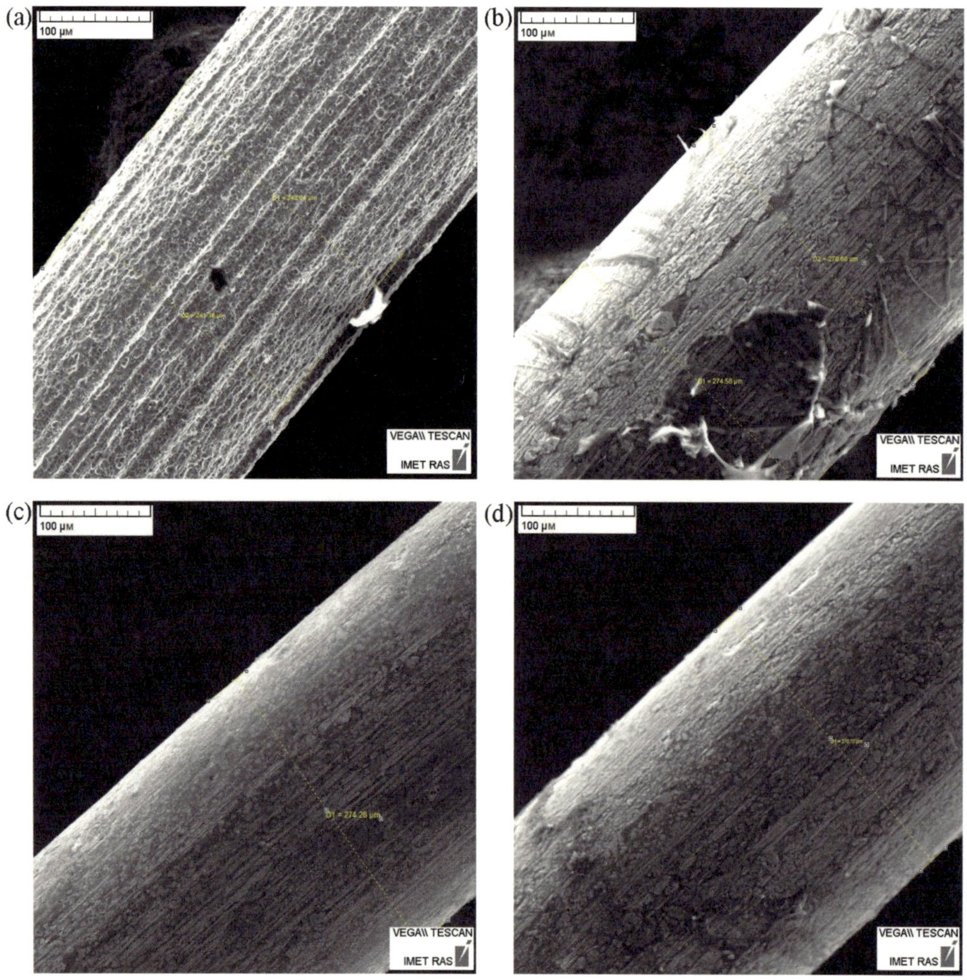

Figure 9. SEM image of the untreated NiTi sample surface after 2 years of immersion in various solutions: (a) buffer at pH 1.68 (Solution 1; **Table 2**); (b) buffer at pH 4.01 (Solution 3, **Table 2**); (c) 0.9 wt% NaCl (Solution 4, **Table 2**); and (d) HCl (pH 1.56).

In the case with solutions of salts of organic acids (pH 1.68–4.01), the dissolution of nickel and titanium is explained because both elements react with them [42]. In addition, regions of selective dissolution of metals were revealed, titanium in acidic media and nickel in acidic and neutral media [48]. However, nickel is subjected to corrosion damage in oxidizing media, and titanium is considered as satisfactorily stable (has a pitting corrosion potential almost two orders of magnitude higher than that for nickel in NaCl solution) [46, 47]. In addition, the investigation of the corrosion mechanism of microstructural nitinol in chlorine-containing solutions showed that the corrosive surface demonstrated a low content of nickel, and the result of the process was the Ni ion release into solution and the formation of titanium oxide in the damaged region because titanium remained at the surface and reacted with dissolved oxygen [12]. The region between volume NiTi and surface TiO_2 contained layers of Ni_3Ti and Ni_4Ti and clusters of pure nickel [1, 41, 11], the formation of which was caused also by the formation of the titanium oxide layer: nickel atoms are released from the Ni-Ti interatomic bond almost at room temperature when the thin oxide layer spontaneously forms at the nitinol surface, and nickel may be released into the solution. However, published data of the titanium ion release at the corrosion of microstructural nitinol were not found. Therefore, the presence of titanium ions in the NaCl solution turned out to be unexpected. Here, we suggest that this is related to the nanostructure of nitinol: nanograins failed during the "washout" of nickel from them and released a greater amount of titanium into the solution at the beginning [42]. Though titanium is not considered to be toxic for humans even in amounts much greater than those obtained here and its concentration in the solution is much less than the nickel content, it can be noted, however, that the nanostructural properties doubly affect the corrosion resistance of nitinol.

We observed that the significant retardation of the nickel ion release (and insignificant concentration as a whole) and the absence of titanium ion release form nanostructural nitinol samples in weakly acidic and neutral solutions after mechanical polishing. Traces of pitting corrosion on the surface of polished samples are visible only after 2 years of holding in the most acidic medium (**Figure 10**); wires held in the remaining media look intact equally. This leads to a conclusion of the occurrence of a strong and homogeneous protective surface layer

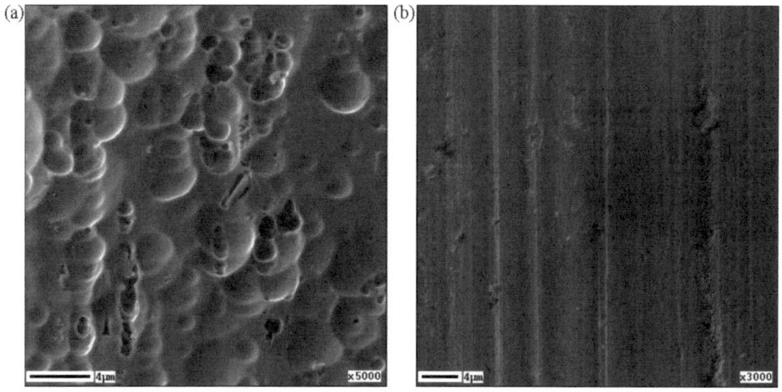

Figure 10. SEM image of the polished wire surface after holding in (a) solution with pH 1.68 and (b) remaining media.

of titanium oxide, which serves as a barrier to the release of nickel into the medium, and the high corrosion resistance of investigated nanostructural nitinol.

According to the literature, with increasing holding period of samples in the chlorine-containing solution, the thickness of the oxide layer increases [12, 14]. The investigation of untreated nitinol wires after 4 months of holding in the salt solution showed that their breakdown potential of the passive film changed from 200 to 800 mV, which the authors related to a decrease in the Ni content on the surface [17].

In this chapter, the surface composition of TiNi wires is practically invariable in the case with untreated and annealed samples after holding in solutions. Dark carbon-containing spots with insignificant nickel content and bright areas of titanium oxide with nickel inclusions, as before, alternate with each other, and the total analysis over an area of 100 × 100 μm² demonstrate the invariably high carbon content at a depth greater than 1 μm.

The greatest depth of the surface oxide layer, approximately 25 nm, is observed for polished wires after holding in neutral solutions (**Table 4**). At the same time, the titanium distribution in the elementary and bound states after the NaCl medium was determined, and elementary titanium was not revealed to a depth of 17 nm (**Figure 11**). The nickel contents at the surface

Solution pH	1.68	3.56	4.01	6.31/7.36	9.18
Oxide layer thickness/nm	8–13	13–17	15–20	23–28	≈10
Nickel surface concentration/wt%	8	5	3	0	3

Table 4. Surface composition of polished NiTi samples after immersion in various solutions for 2 years.

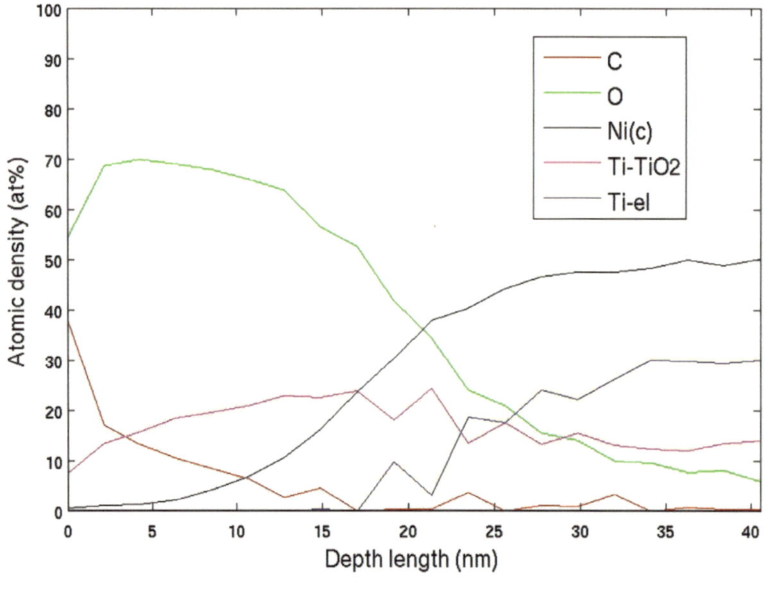

Figure 11. Composition of the polished NiTi wire surface after immersion for 2 years in 0.9 wt% NaCl (Solution 4, Table 2) as determined by Auger spectroscopy.

itself are shown in Table 4, in the NaCl and artificial plasma comes to a plateau at a depth of ≈20 nm, whereas in the remaining media, at about 10 nm. The oxygen concentration in the surface layer in all samples traverses the maximum at ≈60–70 wt% at a depth of 2.5–7 nm. At the same time, the carbon content on these dependences drops abruptly from a fairly high one at a depth to 2.5 nm, which can be related to the presence of outer mechanical contaminations at the surface. Release of metal ions into the solution with pH 9.18 was not observed completely, and this explains the identity of compositions before and after immersion.

Thus, the mechanically polished surface held in the chlorine-containing medium really looks protected from corrosion, including under dynamic conditions, even after subsequent annealing. However, the contact of nickel with the surrounding physiological environment nevertheless is possible that it demands to consider an individual susceptibility of patients.

The biocompatibility of the nanostructured nitinol was measured *in vitro* using standard test systems: the cultures of myofibroblasts from human peripheral vessels and human bone marrow mesenchymal stromal cells (MSC) were used as standard cell models. The myofibroblasts were isolated from cut peripheral veins [49] and grown in the DMEM medium (Biolot, Russia) with the addition of 10% fetal calf serum (Gibco, United States), 40 μg/mL of gentamicin at 37°C, and 5% carbon dioxide in a CO_2 incubator (Binder, Germany). The MSC (Biolot, Russia) were grown in the alpha-MEM medium (Sigma, United States) under the same conditions.

Samples of materials were placed into the wells of a 6-well plate (Greiner, Germany). Then cells were inoculated on the sample surface (5×10^3 cells per cm^2) and cultured for 5 days. To determine the numbers of vital and dead cells, the cells growing on the material surface were stained with fluorescent dyes—acridine orange (Sigma, USA, 1 μg/mL) and propidium iodide (Sigma, USA, 1 μg/mL). Acridine orange stains both vital and dead cells, while propidium iodide stains only dead cells. After that, the samples were incubated for 10 min at 37°C [50]. Then the samples were examined under a DM 6000 fluorescence microscope (Leica, Germany). For the assay, at least 500 cells on the sample surface were counted. In the case of myofibroblasts and MSC, the percentages of vital cells for NiTi were 91 ± 3 and 95 ± 1, respectively. Thus, the material samples used in the study did not have a short-term toxic effect on the cells that overgrew its surfaces *de novo*.

The mitotic activity of the cells was assessed considering the mitotic index of the cells in the logarithmic growth phase (the 3rd day after inoculation). The number of mitotic cells was determined by fluorescence microscopy using the vital staining with the Hoechst 33342 fluorescent dye (Sigma, USA). The mitotic cells were identified based on the distribution of chromatin inherent in the prophase, metaphase, anaphase, and telophase. At least 500 cells on the sample surface were counted for the assay. The calculated MI value [51] for the cells growing on the NiTi surface was 3.1% for the myofibroblast culture and 1.8% for the MSC culture.

After 5 days of culturing, the morphological analysis of the myofibroblasts and MSC on the surface of material showed that myofibroblasts occupy only ≈75% of the NiTi surface accessible for the growth, while MSC occupy ≈50% of the accessible NiTi surface and so no monolayer is formed for either myofibroblasts or MSC.

This means that toxicity of samples was not revealed, but the cell reaction was not the best possible.

Thus it is shown that the use of nanostructured nitinol, obtained and investigated in this work, in medicine as a material for noninvasive implants is prospective compared with microstructured material, but formation of a corrosion-resistant and biocompatible surface layer in future still seems desirable.

Investigated nanostructural material is currently used to manufacture the noninvasive medical implants—stents—successfully applied to restore respiratory and urinary systems and esophageal and intestinal patency (**Figure 12**).

In connection with the presence of increased corrosion resistance, strength and plasticity, shape memory effect, controlled by the composition and heat treatment phase transformations, this

(a)
(b)
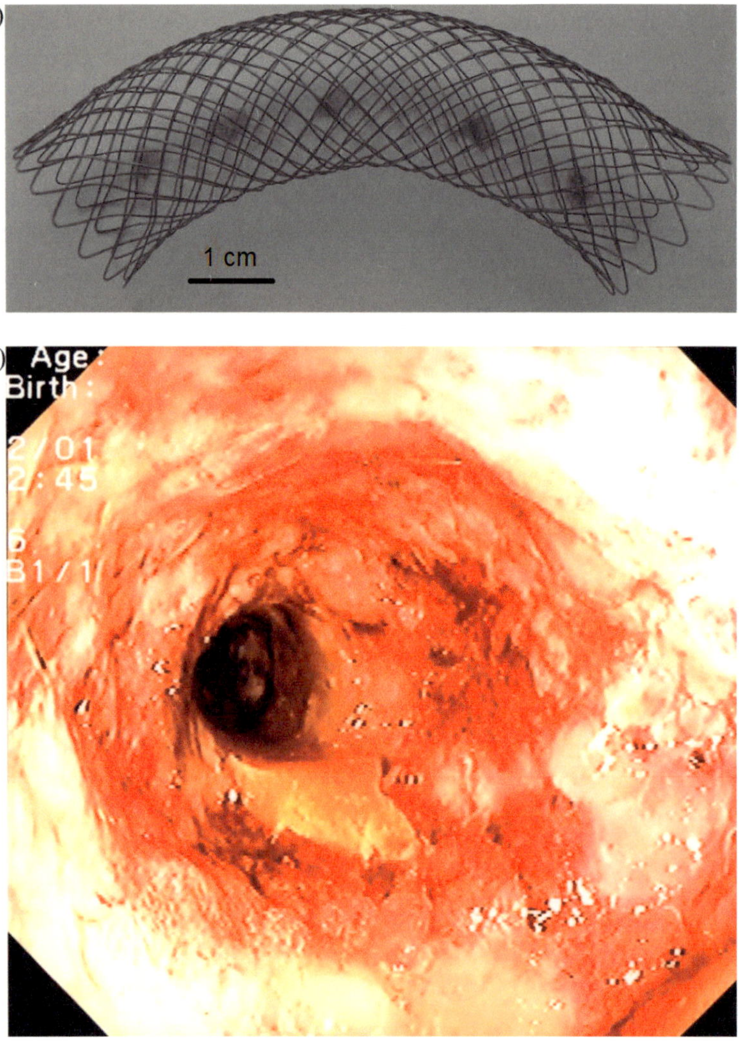

Figure 12. Patency restoration process: (a) stent applied to restore esophageal and intestinal patency, (b) example of the restoration of patency in the case of large intestine cancer via the implantation of a stent (operation and photo of the Blokhin Russian Cancer Research Center, Russian Academy of Medical Sciences).

new nanostructural material can also be used to create other medical products and instruments, as well as can be used in other areas that require the use of smart materials.

3. Conclusions

New nanostructural NiTi alloy that can be used in medical applications was obtained and investigated.

The material base is the B2 phase in the form of nanofibers with a diameter of 30–70 nm, which are elongated along the wire axis. Inclusions of Ti_2Ni intermetallides are also observed, the distribution and the size of which are not affected by annealing at 450°C for 15 min.

The nickel release from investigated nanostructural nitinol is less in comparison with data for microstructural nitinol in a solution of any acidity. Although earlier in the literature, there were no similar results. Dissolution in the alkali medium is absent.

Mechanical treatment increases nitinol corrosion resistance, and thermal treatment decreases it, and the nanostructure possibly serves as the cause of the titanium ion presence in the solution of any acidity with unpolished samples. A significant retardation of the nickel ion release (and insignificant concentration as a whole) and the absence of titanium ion release in the weakly acidic and neutral solutions with polished samples are observed.

The mechanical treatment of the surface results in the formation of a thin layer of titanium oxide, decreases the roughness and number of defects but does not completely remove nickel from the surface. The nickel-free surface with the 25-nm titanium oxide protective layer is attained by holding in a neutral 0.9 wt% solution of sodium chloride. Such a layer must be the barrier to nickel release into the medium even under dynamic conditions.

A simultaneous 7–11% increase in strength and plasticity in comparison with microstructural nitinol was attained. The presence of the shape memory effect and the absence of an uncontrolled change in the temperature of phase transformations in connection with nanostructuring were noted.

Toxicity of samples has not been revealed. But no cell monolayer is formed on the surface of nanostructural alloy after 5 days of culturing of the myofibroblasts and mesenchymal stromal cells.

Nanostructured nitinol usage in medicine as a material for noninvasive implants is prospective compared with microstructured material but formation of corrosion-resistant and biocompatible surface layer still seems desirable.

Acknowledgements

This work was supported by the RFBR № 15-33-70006 «mol_a_mos» and 14-29-10208 «ofi_m». The authors wish to thank Golberg MA, Dyomin K Yu, Mikhailova AB, Volchenkova VA, and Kargin Yu F for their help in sample analysis.

Author details

Elena O. Nasakina*, Mikhail A. Sevostyanov, Alexander S. Baikin, Alexey V. Seryogin, Sergey V. Konushkin, Konstantin V. Sergienko, Alexander V. Leonov and Alexey G. Kolmakov

*Address all correspondence to: nacakina@mail.ru

A.A. Baikov Institute of Metallurgy and Material, Science, Russian Academy of Sciences, Moscow, Russia

References

[1] Shabalovskaya S. On the nature of the biocompatibility and medical applications of NiTi shape memory and superelastic alloys. Bio-Medical Materials and Engineering. 1996;**6**:267-289. DOI: 10.3233/BME-1996-6405

[2] Gyunter VO, Khodorenko VN, Yasenchuk YF, Chekalkin TL. Nikelid titana. Meditsinskii material novogo pokoleniya (Titanium Nickelide. Medical Material of New Generation). Tomsk: MITs; 2006. p. 296

[3] Dotter CT, Buschmann PAC, McKinney MK, Rosch J. Transluminal expandable nitinol coil stent grafting: Preliminary report. Radiology. 1983;**147**:259-260. DOI: 10.1007/s00330-003-2022-5

[4] Zabolotnyi VT, Belousov OK, Palii NA, Goncharenko BA, Armaderova EA, Sevost'yanov MA. Materials science of the production, treatment and properties of titanium nickelide for application in endovascular surgery. Russian Metallurgy (Metally). 2011;**3**:437-448. DOI: 10.1134/S003602951105017X

[5] Stoeckel D. Nitinol medical devices and implants. Minimally Invasive Therapy & Allied Technologies. 2000;**9**:81-88

[6] Bose A, Hartmann M, Henkes HA. Novel. Self-expanding nitinol stent in medically refractory intracranial atherosclerotic stenosis: Wingspan study. Stroke. 2007;**38**:1531-1537. DOI: 10.1161/STROKEAHA.106.477711

[7] Xiaoying Lu, Xiang Bao, Yan Huang, Yinghua Qu, Huiqin Lu, Zuhong Lu. Mechanisms of cytotoxicity of nickel ions based on gene expression profiles. Biomaterials. 2009;**30**:141-148. DOI:10.1016/j.biomaterials.2008.09.011

[8] Nasakina EO, Sevostyanov MA, Goncharenko BA, Leonova YO, Kolmakov AG, Zabolotny VT. Metody issledovaniya I povysheniya korrozionnoy stoykosty meditsynskogo splava s effektom pamiaty formy NiTi. Issledovanie korrosionnoy stoykosty I biosovmestimosty nitinola [Methods of corrosion resistance research and increase of medical alloy with shape memory effect NiTi. Research of corrosion resistance and biocompatibility of nitinol]. Perspektivnie materialy [Advanced materials]. 2014;**7**:37-49

[9] Nasakina EO, Sevostyanov MA, Goncharenko BA, Leonova YO, Kolmakov AG, Zabolotny VT. Metody issledovaniya I povysheniya korrozionnoy stoykosty meditsynskogo splava s effektom pamiaty formy NiTi. Sposoby izmeneniya korrozionnoy stoykosti nitinola [Methods of corrosion resistance research and increase of medical alloy with shape memory effect NiTi. Ways of a nitinol corrosion resistance change]. Perspektivnie materialy [Advanced materials]. 2014;**9**:19-33

[10] Wever DJ, Veldhuizen AG, de Vries J, Busscher HJ, Uges DRA, van Horn JR. Electrochemical and surface characterization of a nickel-titanium alloy. Biomaterials. 1998;**19**: 761-769. DOI: 10.1016/S0142-9612(97)00210-X

[11] Shabalovskaya SA., Tian He, Anderegg JW, Schryvers DU, Carroll WU, Van Humbeeck J. The influence of surface oxides on the distribution and release of nickel from Nitinol wires. Biomaterials. 2009;**30**(4):468-477. DOI: 10.1016/j.biomaterials.2008.10.014

[12] Hu T, Chu C, Xin Y, Wu S, Yeung KWK, Chu PK. Corrosion products and mechanism on NiTi shape memory alloy in physiological environment. Journal of Materials Research. 2010;**25**:350-358. DOI: 10.1557/JMR.2010.0051

[13] Trepanier C, Venugopalan R, Messer R, Zimmerman J, Pelton AR. Effect of passivation treatments on nickel release from nitinol. In: Proceedings of the 6th World Biomaterials Congress Transactions; 2000; Honolulu. Minneapolis: Society for Biomaterials; 2000. p. 1043

[14] Barrett RD, Bishara SE, Quinn JK. Biodegradation of orthodontic appliances: Part I. Biodegradation of nickel and chromium in vitro. American Journal of Orthodontics and Dentofacial Orthopedics. 1993;**103**:8-14. DOI:10.1016/0889-5406(93)70098-9

[15] Her-Hsiung Huang, Yu-Hui Chiu, Tzu-Hsin Lee, Shih-Ching Wu, Hui-Wen Yang, Kuo-Hsiung Su, Chii-Chih Hsu. Ion release from NiTi orthodontic wires in artificial saliva with various acidities. Biomaterials. 2003;**24**:3585-3592. DOI: 10.1016/S0142-9612(03)00188-1

[16] Uo M, Watari F, Yokoyama A, Matsuno H, Kawasaki T. Dissolution of nickel and tissue response observed by Xray scanning analytical microscopy. Biomaterials. 1999;**20**:747-755. DOI: 10.1016/S0142-9612(98)00224-5

[17] Clarke B, Kingshott P, Hou X, Rochev Y, Gorelov A, Carroll W. Effect of nitinol wire surface properties on albumin adsorption. Acta Biomaterialia. 2007;**3**:103-111. DOI: 10.1016/j.actbio.2006.07.006

[18] Her-Hsiung Huang. Surface characterizations and corrosion resistance of nickel–titanium orthodontic archwires in artificial saliva of various degrees of acidity. Journal of Biomedical Materials Research Part A. 2005;**74A**:629-639. DOI: 10.1002/jbm.a.30340

[19] Shabalovskaya SA, Anderegg J, Wataha J, Adler P, Cunnick J. Effects of surface chemistry on biocompatibility of NiTi. In: Proceedings of the International Conference on Shape Memory and Superelastic Technologies; October 3-7, 2004; Kurhaus Baden-Baden. Materials Park: ASM International; 2004. pp. 367-375

[20] Her-Hsiung Huang. Variation in corrosion resistance of nickel-titanium wires from different manufacturers. Angle Orthodontist. 2005;**75**:661-665

[21] Trepanier C, Tabizian M, Yahia LH, Bilodeau L, Piron DL. Effect of modification of oxide layer on NiTi stent corrosion resistance. Journal of Biomedical Materials Research. 1998;43:433-440

[22] Trepanier C, Fino J, Zhu L, Pelton AR. Corrosion resistance of oxidized nitinol. In: Proceedings of the International Conference on Shape Memory and Superelastic Technologies. SMST-2003; 5-8 May 2003; California. California: SMST Soc; 2004. pp. 267-276

[23] Farrukh MA, editor. Functionalized Nanomaterials. Rijeka: InTech; 2016. pp. 172. DOI: 10.5772/63186

[24] Roldugin VI, Fedotov MA, Folmanis GE, Kovalenko LV, Tananaev IG. Formation of aqueous colloidal solutions of selenium and silicon by laser ablation. Doklady Physical Chemistry. 2015;**463**:161-164. DOI: 10.1134/S0012501615070064

[25] Leon Mishnaevsky Jr., Levashov E, Ruslan Z, Valiev, Javier Segurado, Ilchat Sabirov, Nariman Enikeev, Sergey Prokoshkin, Andrey V Solov'yov, Andrey Korotitskiy, Elazar Gutmanas, Irene Gotman, Eugen Rabkin, Sergey Psakh'e, Luděk Dluhoš, Marc Seefeldt, Alexey Smolin. Nanostructured titanium-based materials for medical implants: Modeling and development. Materials Science and Engineering R. 2014;**81**:1-19. DOI: 10.1016/j.mser.2014.04.002

[26] Amirhanova NA, Valiev RZ, Adasheva SL, Prokof'ev EA. Issledovanie korrozionnyh i jelektrohimicheskih svojstv splavov na osnove nikelida titana v krupnozernistom i ul'tramelkozernistom sostojanijah [Investigation of corrosion and electrochemical properties of alloys based on nickel titanium in coarse and ultrafine states]. Journal of the Ufa State Aviation Technical University. 2006;**7**(1):143-146

[27] Korotin DM, Bartkowski S, Kurmaev EZ, Cholakh SO, Müller M, Neumann M, Gunderov D, Valiev RZ, Cholakh SO. Arsenic contamination of coarse-grained and nanostructured nitinol surfaces induced by chemical treatment in hydrofluoric acid. Journal of Biomedical Materials Research Part B: Applied Biomaterials. 2012;**100B**(7):1812-1816. DOI: 10.1002/jbm.b.32748

[28] Zheng CY, Nie FL, Zheng YF, Cheng Y, Wei SC, Valiev RZ. Enhanced in vitro biocompatibility of ultrafine-grained titanium with hierarchical porous surface. Applied Surface Science. 2011;**257**(13):5634-5640. DOI: 10.1016/j.apsusc.2011.01.062

[29] Zheng CY, Nie FL, Zheng YF, Cheng Y, Wei CS, Liqun Ruan, Valiev RZ. Enhanced corrosion resistance and cellular behavior of ultrafine-grained biomedical NiTi alloy with a novel SrO-SiO_2-TiO_2 sol-gel coating. Applied Surface Science. 2011;**257**(13):5913-5918. DOI: 10.1016/j.apsusc.2011.02.006

[30] Zhang XN. Biomechanical and biocorrosion properties of nanostructured titanium. Advanced Materials Research. 2007;**29-30**:51-54. DOI: 10.4028/www.scientific.net/AMR.29-30.51

[31] Aslan Ahadi, Qingping Sun. Effects of grain size on the rate-dependent thermomechanical responses of nanostructured superelastic NiTi. Acta Materialia. 2014;**76**:186-197. DOI: 10.1016/j.actamat.2014.05.007

[32] Delville R, Malard B, Pilch J, Sittner P, Schryvers D. Transmission electron microscopy study of dislocation slip activity during superelastic cycling of NiTi. International Journal of Plasticity. 2010;**27**:282-297. DOI: 10.1016/j.ijplas.2010.05.005

[33] Shu-yong Jiang, Ya-nan Zhao, Yan-qiu Zhang, Ming Tang, Chun-feng LI. Equal channel angular extrusion of NiTi shape memory alloy tube. Transactions of Nonferrous Metals Society of China. 2013;**23**:2021-2028. DOI: 10.1016/S1003-6326(13)62691-6

[34] Zhilyaev AP, Langdon TG. Using high-pressure torsion for metal processing: Fundamentals and applications. Progress in Materials Science. 2008;**53**:893-979. DOI: 10.1016/j.pmatsci.2008.03.002

[35] Stolyarov VV. Deformability and nanostructuring of TiNi shape-memory alloys during electroplastic rolling. Materials Science and Engineering A. 2009;**503**:18-20. DOI: 10.1016/j.msea.2008.01.094

[36] Liu HS, Mishnaevsky Jr L. Martensitic transformations in nanostructured nitinol: Finite element modeling of grain size and distribution effects. Computational Materials Science. 2013;**76**:27-36. DOI: 10.1016/j.commatsci.2012.11.032

[37] Adharapurapu RR, Vecchio KS. Effects of aging and cooling rate on the transformation of nanostructured Ti-50.8Ni. Journal of Alloys and Compounds. 2017;**693**:150-163. DOI: 10.1016/j.jallcom.2016.09.112

[38] Vojtech D, Michalcová A, Capek J, Marek I, Dragounová L. Structural and mechanical stability of the nano-crystalline Ni-Ti (50.9 at.% Ni) shape memory alloy during short-term heat treatments. Intermetallics. 2014;**49**:7-13. DOI: 10.1016/j.intermet.2013.12.013

[39] Divya Rani VV, Vinoth-Kumar L, Anitha VC, Manzoor K, Deepthy M, Nair Shantikumar V. Osteointegration of titanium implant is sensitive to specific nanostructure morphology. Acta Biomaterialia. 2012;**8**:1976-1989. DOI: 10.1016/j.actbio.2012.01.021

[40] Shishkovsky I, Morozov Yu., Smurov I. Nanofractal surface structure under laser sintering of titanium and nitinol for bone tissue engineering. Applied Surface Science. 2007;**254**:1145-1149. DOI:10.1016/j.apsusc.2007.09.021

[41] Zhu L, Trepanier C, Fino J, Pelton AR. Oxidation of nitinol and its effect on corrosion resistance. In: Proceedings of the Conference on ASM Matererials. Proceedings for Medical Device; 2003; Anaheim. Materials Park, Cleveland: ASM International; 2004. pp. 8-10

[42] Nasakina EO, Baikin AS, Sevost'yanov MA, Kolmakov AG, Zabolotnyi VT, Solntsev KA. Properties of nanostructured titanium nickelide and composite based on it. Theoretical Foundations of Chemical Engineering. 2014;**48**(4):477-486. DOI: 10.1134/S0040579514040071

[43] Tomic S, Rudolf R, Brun ko M, Anžel I, Savic V, Olic M. Response of monocytederived dendritic cells to rapidly solidified nickeltitanium ribbons with shape memory properties. European Cells & Materials. 2012;**23**:58-81. DOI: 10.22203/eCM.v023a05

[44] Venugopalan R, Trepanier C. Assessing the corrosion behavior of nitinol for minimally-invasive device design. Minimally Invasive Therapy & Allied Technologies. 2000;**9**:67-74

[45] Sun EX, Fine S, Nowak WB. Electrochemical behavior of nitinol alloy in ringer's solution. Journal of Materials Science: Materials in Medicine. 2002;**13**:959-964. DOI: 10.1023/A:1019812729884

[46] Zhuk NP. Kurs korrozii i zashhity metallov [The course of corrosion and protection of metals]. Moscow: Metallurgy; 1976

[47] Ulig GG, Revi GU. Korrozija i bor'ba s nej. Vvedenie v korrozionnuju nauku i tehniku [Corrosion and its prevetion]. Leningrad: Chemestry; 1989

[48] Shcherbakov AI, Kasatkina IV, Zalavutdinov RK. Ustojchivost' splava «pamjati» TiNi k selektivnoj korrozii [Resistance of "Memory" alloy TiNi to selective corrosion]. Corrosion: Materials and Protection. 2007;**12**:14-16

[49] Akatov VC, Fadeeva IC, Chekanov AV, Solov'yev VV. Rol' kletok retsipiyenta v mekhanizme patologicheskoy kal'tsifikatsii transplantatov klapanov serdtsa i sosudov [Role of the recipient cells in the mechanism of transplants pathological calcification of heart valves and blood vessels]. Biofizika. 2010;**55:**937-942

[50] Garmash SA, Smirnova VS, Karp OE, Usacheva AM, Berezhnov AV, Ivanov VE, Chernikov AV, Bruskov VI, Gudkov SV. Pro-oxidative, genotoxic and cytotoxic properties of uranyl ions. Journal of Environmental Radioactivity. 2014;**127**:163-170. DOI: 10.1016/j.jenvrad.2012.12.009

[51] Permyakov SE, Knyazeva EL, Khasanova LM, Fadeev RS, Zhadan AP, Roche-Hakansson H, Håkansson AP, Akatov VS, Permyakov EA. Oleic acid is a key cytotoxic component of HAMLET-like complexes. Biological Chemistry. 2012;**393**:85-92

Chapter 5

Development of Faster SMA Actuators

Atta Muhammad Nizamani, Jawaid Daudpoto and
Muhammad Ali Nizamani

Additional information is available at the end of the chapter

http://dx.doi.org/10.5772/intechopen.69868

Abstract

Large cycle time, resulted from slow cooling, is the core hindrance to the wide spread applications of shape memory alloys (SMAs) as actuators. This chapter discusses a novel cooling technique to decrease the cycle time of SMAs. Under this technique, the SMA actuator of 0.15 mm diameter was run through a grease-filled Polytetrafluoroethylene (PTFE) tube of 0.5 mm outside diameter. Later, same tests were repeated with oil filled PTFE tube. The test results conducted in ambient air were used as standard for comparison. The actuation current in ambient air was set at 210, 310 and 410 mA. While testing with heat sink, i.e. grease and oil, the SMA was heated with 210, 310, 410, 500, 615 and 720 mA currents for 1 and 2 seconds, whereas the SMA was heated for 1 second only with 810 mA current. It was found that the grease cooling reduced the cooling time up to 30% and oil cooling by 20%, as compared to the ambient air-cooling time. However, the grease-cooled actuators had shown less strain, and their response was non-linear at many instances. Heat loss to the sinks resulted to more power consumption than that in ambient air cooling for equivalent amount of strain.

Keywords: SMA actuators, long cycle time, SMA cooling, Teflon tubing

1. Introduction

The need for miniaturization and lighter systems has resulted in the development of smart actuators, which are compact in size and lighter in weight. The shape memory alloys (SMAs), also called the smart alloys, were discovered by Arne Ölander in 1932 [1]. Various properties of SMAs, like high work output as compared to other conventional actuators, silent, clean and spark-free operation, design simplicity and easy miniaturization, have attracted researchers and engineers to use them as actuators in several applications [2].

The operation of SMA actuators is considered to be simple. SMA actuators in many applications are in the form of a thin wire and their diameters ranging from 0.025 to 0.51 mm [3]. The motion created in SMAs is due to a molecular rearrangement in their crystalline structure when phase transformation takes place, as shown in **Figure 1**.

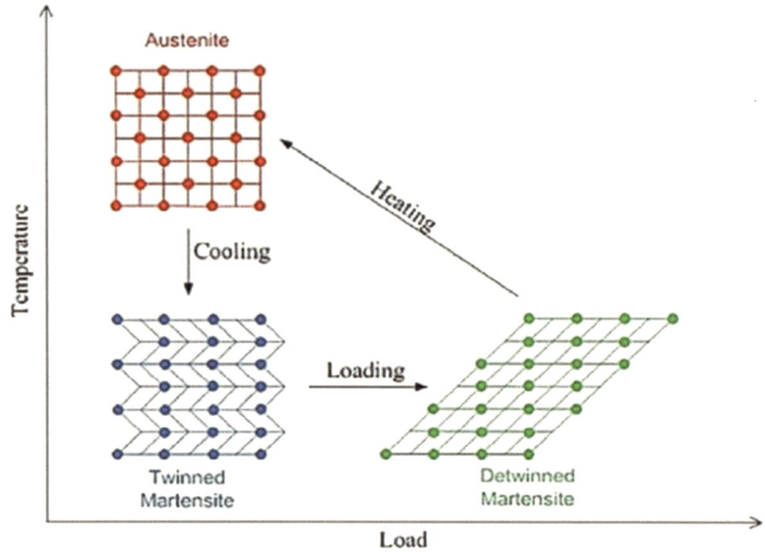

Figure 1. Phase transformation in SMAs [4].

The phase transformation takes place at certain temperatures called phase transformation temperature. The phase transformation temperature is determined by the composition and heat treatment methods applied to the SMAs. Low temperature phase is known as martensite, whereas, heating results in transformation to austenite phase. SMAs are relatively soft in Martensite phase and possess smaller value of Young's modulus, whereas they are hard in Austenite phase with higher value of Young's modulus.

2. SMA design challenges

The design challenges in SMA-based systems are to succeed over the limitations incorporated with SMAs. Some of the SMA limitations are relatively small usable strain, low operational frequency, low controllability, low accuracy and low-energy efficiency. However, low operational frequency resulted from large cycle time is widely reported in literature as core hindrance to wide spread applications of the SMAs [5–7]. The operational frequency of SMAs is given by Eq. (1) (working frequency of SMAs).

$$f_w = \frac{1}{t_h + t_c} \tag{1}$$

http://dx.doi.org/10.5772/intechopen.69868

Where

f_w = Working frequency of SMAs

t_h = Heating time required by SMAs

t_c = Cooling time required by SMAs

The cycle time in SMAs is defined as the total time required by the SMAs to contract and expand, hence completing its full cycle. Cycle time is the algebraic sum of heating time (i.e. contraction time) and cooling time (i.e. expansion time). The heating time of the SMAs can be easily reduced by increasing the magnitude of actuation current, whereas the cooling rate is redistricted by the rate of heat transfer rate to the environment surrounding the SMA specimen. The actuator response time also depends upon the size and shape of the actuators. The SMA actuators with smaller diameter cool faster as compared to those having larger diameters. However, this will affect the loading capacity of SMA actuators. **Figure 2** is developed by using the technical data provided by Dynalloy, Inc., the SMA manufacturer [3]. It shows the relationship between the loading capacity of SMA actuators and cooling time required by SMA actuators of various diameters in ambient air at room temperature. However, the time required to restore final 0.5% strain is not considered.

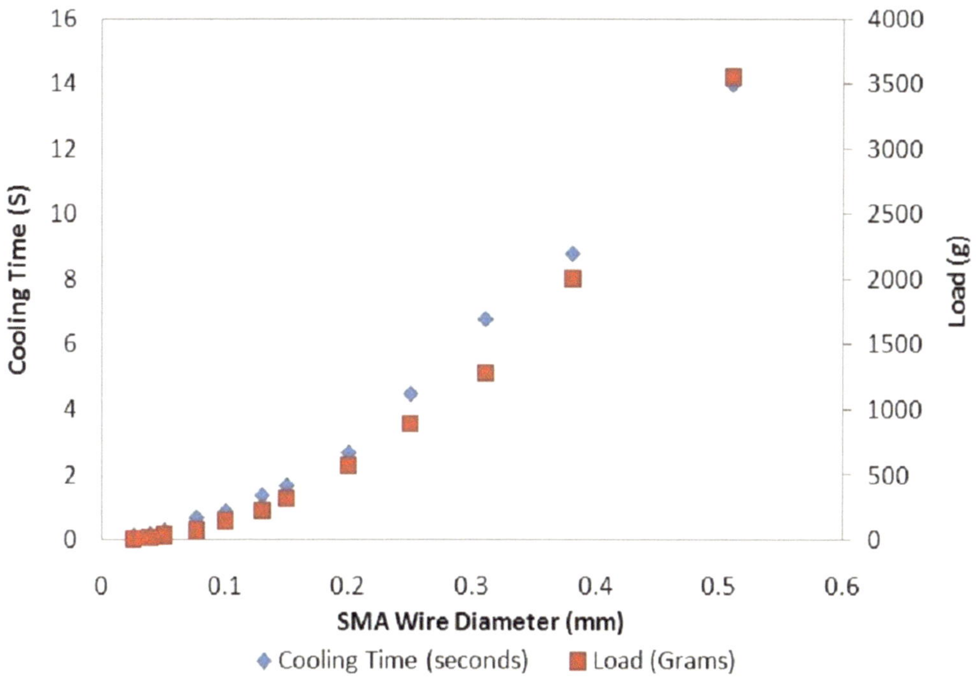

Figure 2. Relationship between wire diameter, cooling time and safe load for SMA wire actuators.

From **Figure 2**, it can be observed that the cooling time required by the smaller diameter SMAs is relatively small. However, the smaller diameter SMAs can carry little load as compared to those carried by SMAs having larger diameter. Therefore, the need is to develop the faster cooling mechanisms for SMAs.

Large cycle time is core hindrance to the SMA applications in several areas, including automobiles, robotics, biomedical, etc.

3. Practical applications of faster SMA actuators

Around 200 actuation tasks are performed in a family car [8]. The actuators required to operate several functions in a car are categorized as (a) low-power actuators for comfort and aesthetic aspects, (b) actuators having high power to operate various control mechanisms of vehicles and (c) the actuators with high frequency to control engine performance [9]. The SMA actuators are being used in first category of actuators, as applied by Mercedes, BMW, General Motors (GM), Hyundai, Ford, Porsche and Volkswagen (VW) to actuate lumbar support in the car seats for passenger comfort [10]. The SMAs are applicable in second category of actuators; however, these are still to be studied for this category, whereas due to low working frequency, resulting from large cooling time, the SMA actuators are either not suitable or less suitable to be applied against third category of actuators.

The SMAs due to their ability to produce linear motion, high power to weight ratio, light weight and compact size are widely being used in several areas of robotics. Some common SMA applications in robotics are robotic grippers [11], human organ orthotics like a foot ankle orthotics [12], rehabilitation robots like a wearable elbow exoskeleton [13] and bio-inspired fish-like robots like i-tuna and dragon fly robots [14, 15]. However, large cooling time is reported to be the core limitation of SMAs in the field of robotics [8, 12, 16, 17].

The biocompatibility of SMAs has made them feasible to be applied in several applications related to biomedical sciences. SMAs are being used in neurology, cardiology and interventional radiology in different ways [18]. SMA-actuated surgical needles and catheters, artificial heart, artificial head and drug-delivery valves are some of the most prominent SMA applications in area of biomedical [19–22]. However, low working frequency resulted from large cooling time is reported to be the serious issue [19, 22, 23].

The SMAs due to attractive morphing properties and ability to withstand dynamic loads are used in several aerospace applications. Some of the noticeable aerospace applications of SMAs are in wing morphing, for example, the DARPA project for development of smart wings [24], SAMPSON project to enhance SMA applications in aircraft engine nozzles at inlets to obtain flying benefits according to flight conditions [25] and reconfigurable engine nozzle fan chevron by Boeing [26]. SMAs are equally applicable in various space research applications, for example, folding and unfolding mechanisms for solar panels on Hubble telescope [27], actuation of valves and apertures on board of Rosetta mission (2004) and Pathfinder–Sojourner Mission (1997) [28]. Their applications can still be enhanced to various actuation applications in spacecrafts, either manned or unmanned [29]. However, the limited working frequency of SMAs resulting from large cooling time is core issue in many aerospace engineering applications [30, 31].

4. Literature review

Several studies have been conducted to reduce the cooling time required by SMAs to regain martensite phase; however, no any unique method has been proposed. In the following passages, a review of such studies is presented.

Loh et al. [32] used silicone grease as heat sink. The authors filled the grease in a metal tube of outside diameter 0.8 mm and inserted 600 mm long and 0.3 mm diameter SMA actuator in the tube.

In reported method, the heat dissipation from SMA to sink was achieved at 5°C per second. The SMA actuator was heated by electrical current supply of 2 A, with a duty ratio of 0.4. The martensitic phase transformation also depends upon the mechanical load connected to the wire; however, in the reported work, the SMA actuators are made to lift up 3 kg mass. The system had following listed limitations:

i. The metal tube is solid; therefore, it is likely to affect the system's flexibility.

ii. Since the actuator was jacketed around with coolant, the heat loss at 1.75°C per second was observed during the actuation. Heat loss during the actuation will increase the heating time required for phase transformation.

In another study, Loh et al. [33], while developing the SMA-actuated prosthetic hand, used stainless steel (SS) tubes to enhance the heat transfer rate from SMAs to the environment. The authors ran 2 SMA wires, each 500 mm long and 0.3 mm diameter, through the SS tubes. The SMA actuators were actuated with 50–70 V via pulse width module (PWM). In the reported method, the actuation voltage ranging from 50 to 70 V is unsafe for many applications and uneconomic as well.

Taylor and Au [17] developed an SMA-actuated prosthetic hand. The authors applied the forced air-cooling technique for rapid heat transfer from the SMA wire actuators to the surrounding environment using a small fan. The forced air cooling with a fan produced satisfactory results; however, this method is not applicable in the miniature applications where space is a major constraint.

Cheng and Desai [34] applied water circulation through the SMA actuator arrangement for cooling purpose. The SMA spring actuator was enclosed inside a silicone tube of inside diameter 1.98 mm. The reported method resulted in 0.33 Hz frequency. However, in the presented research, the working frequency of 0.33 Hz was achieved in ambient air cooling when the SMA actuator was heated with 410 mA actuation current. The water circulation, as applied in reported research, is likely to increase the system's complexity as a pump will be required to circulate the water through the actuator, also special sealing arrangements are necessary to avoid water leakage out of actuator. Therefore, the reported system will be bulky and not feasible in miniature applications.

Pathak et al. [35], in order to examine the performance of various cooling media, conducted a comparative study. The authors conducted tests on various SMA actuator samples whose diameters were ranging from 0.1524 to 0.508 mm and each SMA actuator sample was 177.8 mm long. The authors tested the SMA actuator samples in still air, forced air convection, mineral oil, thermal grease and water as cooling media. In still air, the authors found that SMA wire diameter had major impact on the cooling time. It was observed that the values for coefficient of heat transfer (h) for 0.1524 mm diameter SMA wire was 153 W m^{-2} K^{-1}, whereas value of h for 0.508 mm diameter SMA wire was 68 W m^{-2} K^{-1}, which is 44% smaller than that of 0.1524 mm diameter SMA wire. In forced convection in air, the authors set the air flow rate at 625 ft min^{-1} using a fan. It was noticed that the value of h for 0.1524 mm diameter SMA wire was nearly 650 W m^{-2} K^{-1}, whereas that of 0.508 mm diameter SMA wire was nearly 400 W m^{-2} K^{-1} which is 40% smaller than h values for 0.1524 mm diameter SMA wire. Since the air was enforced using a small fan which is not feasible for miniature applications, for example, in minimal invasive surgery, a separate controller will be required to vary the fan speed in order to adjust the speed of airflow in the SMA actuator, at the same time, this will lead to more power consumption to operate a fan and a possible controller. Later, the authors tested the SMA actuators jacketed around with mineral oil. The value of h for 0.1524 mm diameter SMA wire was found to be 1000 W m^{-2} K^{-1}, whereas the h value for 0.508 mm diameter SMA wire was nearly 510 W m^{-2} K^{-1}. This shows a reduction of 49% in the h values for 0.508 mm diameter SMA wire as compared to that of 0.1524 mm diameter SMA wire. As compared to forced air convection, the h value for 0.1524 mm diameter SMA wire was improved by 350 W m^{-2} K^{-1}, whereas the h value for 0.508 mm diameter SMA wire was improved by 110 W m^{-2} K^{-1}. However, when compared to still air cooling, the h value increased by 6.5 times for 0.1524 mm diameter SMA wire and 8.9 times for 0.508 mm diameter SMA wire. Although this is a significant improvement in the values of coefficient of heat transfer, the oil cooling is incorporated with certain limitations, which include sealing complexity to avoid seepage of oil, also in certain conditions, the viscosity of oil may provide the resistance to SMA motion. The authors later characterized the SMA actuator in thermal grease. It was found that the value of h for 0.508 mm diameter SMA wire was 55% lower as compared to h value of 0.1524 mm diameter SMA wire when thermal grease was used as heat sink. The value of h was significantly higher in thermal grease as compared to the h values in mineral oil and air. For example, the value of h for 0.1524 mm diameter SMA wire was 4.3 times higher as compared to that in mineral oil and 28 times higher than that in still air for same diameter SMA wire. The cooling with distilled water produced better results and cooled off the SMA actuators in least time as compared to all other cooling media. The water quenching increased the value of h by 1.3 times for 0.1524 mm diameter SMA wire and 1.6 times for 0.508 mm diameter SMA wire as compared to h values in thermal grease. However, water cooling has certain costs. However, the major limitation is the boiling temperature of water, as the temperature will exceed 100°C, the water will start boiling; therefore, it is only applicable when the transformation temperature is lower, for example, in Flexinol® 70°C. For water, some special sealing arrangements are necessary. Also, the water circulation without a pump is difficult.

Tadesse et al. [36] conducted a series of tests on Flexinol® and Biometal® SMA actuators manufactured by Dynolloy, Inc and Toki Corporation, respectively. Each sample had 0.1 mm diameter and 100 mm long. Considering the actuation current limitations, the authors actuated the Flexinol® SMA actuator with 180 mA and Biometal® wire with 200 mA actuation current.

The authors tested the SMA samples in various cooling media including forced air at 0.3 m s^{-1} using a small computer fan, forced air cooling at 4.6 m s^{-1} using a compressor, thermal grease, solid heat sink and water quenching. However, the cooling time in ambient conditions was used as standard to compare the results of other cooling media.

The SMA actuators cooled off in 1.1 second while let to cool in ambient air, whereas forced air cooling at 0.3 m s^{-1} improved the cooling time to 0.7 second. The high-speed air circulation cooled off the SMA actuators in about 0.3 second. Since the high-speed air circulation was achieved using a compressor, it is not practically applicable in miniature and weight conscious SMA systems. Solid heat sink was used in the form of aluminium tubes which were brought in contact with SMA actuators up on the power cut off to cool the SMA elements. The solid heat sink cooled the SMA actuators in 0.41 second. Solid heat sinking as proposed by the authors is not applicable in miniature applications, also it will require a close eye over the power shut down to bring the sink in contact with SMA actuator as soon as power supply is cut off. Water quenching cooled the SMA actuators in 0.2 second, which is significant reduction in the cooling time of SMAs as compared to ambient air cooling. However, the water circulation without a compressor is difficult, whereas water spray with a syringe is not practical in many applications including prosthesis, robotics and minimal invasive surgical applications. The thermal grease cooling was applied by filling it in a copper tube along with SMA actuators. The SMA actuator failed to contract while cooling with thermal grease. The author claimed that the reason for the failure in contraction was the fast dissipation of heat from the SMA actuator, which retained insufficient heat to cause a phase transformation in the actuator. However, in the proposed research, the SMA element was strained up to 4% when actuated with 810 mA actuation current, 410 mA actuation current also generated sufficient strain in the SMA actuator.

Russell and Gorbet [37] developed mobile heat sink with a two-wired differential type configuration. The authors used a 350 mm long SMA wire actuator of 0.3 mm diameter. The mechanism was arranged in such a way that the midpoint of the SMA actuator was anchored to a 6 mm diameter shaft, hence leaving 150 mm long SMA actuator on either sides of shaft. Furthermore, the heat sink, in a strip form, was attached to the shaft at the centre. As the wire contracted upon heating, it caused the shaft to rotate which simultaneously rotated the heat sink, attached to the shaft, towards the hot portion of the SMA wire, whose current supply was just disconnected. The reported SMA cooling mechanism is complex and is not applicable in conditions when smaller lengths of SMA actuator are used.

A comprehensive literature review is given in **Table 1**.

Sr. No.	Author	Cooling Media	Remarks
1	Loh et al. [32]	Silicone grease filled in copper tube	SMA Cooled off at 5°C per second, whereas heat loss at 1.75°C per second was observed which will result in more power consumption.
2	Loh et al. [33]	Stainless steel metal tube	Since the system is actuated with 50–70 V PWM, which is too high. Also, the metal tube, being rigid, will affect the system's flexibility in motion.
3	Taylor and Au [17]	Forced air cooling by a small fan	Although the proposed method is efficient to cool the SMA at faster rate, it is not suitable for miniature applications, as in catheters.

Sr. No.	Author	Cooling Media	Remarks
4	Cheng and Desai [34]	Water circulation	Water circulation produced the satisfactory results. However, the boiling temperature of water is core hindrance, also the water being a low viscous fluid will require special sealing arrangement.
5	Pathak et al. [35]	Still air	The cooling time increases with increase in SMA wire diameter.
		Forced air convection	Forced airflow can be achieved using a fan, which is not practical in miniature applications, like minimal invasive surgery.
		Mineral oil	Mineral oil produced a significant improvement in cooling time, as it was reduced up to 48% as compared ambient air cooling. However, in certain applications, oil sealing may be a problem.
		Thermal grease	Thermal grease reduced the cooling time by 55%. However, heat loss during actuation is a serious issue as reported by [6, 32].
		Water	Cooling with water produced best results amongst all other coolants. However, sealing to avoid leakage is core problem.
6	Tadesse et al. [36]	Forced air using computer fan	Airflow using a fan, which is not practical in miniature applications, like minimal invasive surgery. Also, special control algorithms may be required to control air speed.
		Forced air using compressors	The use of compressor is not applicable in miniature, biomedical weight conscious applications of SMAs.
		Thermal grease	The SMA actuator failed to contract due to rapid heat loss to the coolant i.e. thermal grease.
		Solid heat sink	Aluminium tubes, used as solid heat sink, were brought in contact with SMA actuator as power was cutoff. This will require a close look on power supply so that the heat sink may be brought in contact as soon as the power is cut off, or this will require complex control algorithms.
		Water	Cooling with water produced best results amongst all other coolants. However, sealing to avoid leakage is core problem and boiling temperature of water are core hindrances.
7	Russell and Gorbet [37]	Mobile heat sink with two-wire differential type configuration	The reported mechanism is complex and is not applicable with short SMA wires.

Table 1. Review of various cooling media.

5. Materials and methods

5.1. Experimental setup

A special purpose setup was developed to test the SMA actuators for various parameters. The experimental setup comprised of a laser displacement sensor, a load cell, a k-type thermocouple, a current transducer, an NI Elvis prototyping board, my RIO devices and power supply units to actuate the SMA actuator and to power the sensors. The experimental setup is shown in **Figure 3**.

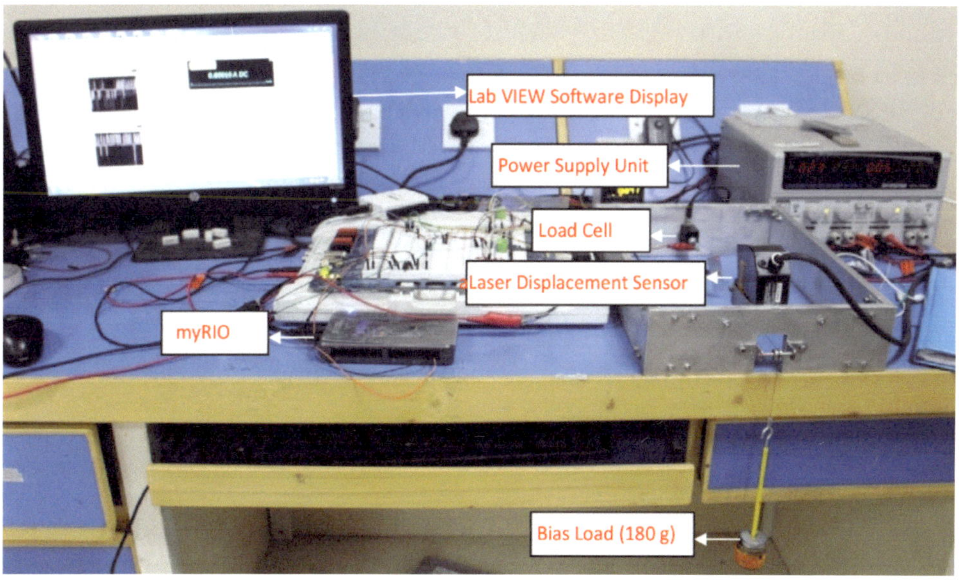

Figure 3. Experimental setup.

5.1.1. Sensors

The special-purpose experimental setup consisted of different sensors, including a laser displacement sensor to determine the contraction (i.e. strain) in the SMA actuators, a load sensor to determine the force exerted by SMA in lifting up the dead weight, a k-type thermocouple to measure the temperature of actuator and a current transducer to determine the current flow across the SMA actuator.

5.1.1.1. Laser displacement sensor

These are the non-contact sensors used to determine the displacement or deformation in case of SMA actuators. Due to high accuracy, these sensors are preferred over the traditionally used displacement sensors, like proximity sensors. In the presented research, the LK-G157 laser displacement sensor made by Keyence was used to detect any deformation, i.e. strain in the SMA actuators. This laser displacement sensor (LK-G157) is capable to measure the deformation when the object is within the range of 150 mm; however, it can measure a maximum of 40 mm displacement. It is provided with a red semiconductor laser which emits light of wavelength 655 nm.

5.1.1.2. Load cell

The tension produced in the SMA actuator while lifting dead weight was determined using a load sensor. Transducers Kit load cell model mdb 2.5 lb, having capacity to measure 2.5 lb, was used in this research.

5.1.1.3. Thermocouple

A close look at the temperature of the SMA actuator is necessary in order to avoid over heating of the SMA actuator, which can cause a permanent deformation. A fine gauge k-type thermocouple along with a MAX7785 amplifier was used to measure the temperature of the SMA actuators. Since the SMA wire diameter was very fine (i.e. 0.15 mm), the contact loss between SMA and the thermocouple was a serious issue, which is addressed by Ref. [38].

5.1.1.4. Current transducer

The strain in the SMA actuator is proportional to the amount of actuation current. An ACS 712 current transducer having analogue output was used to keep a close look on the amount of current being flowing through the SMA actuator.

5.1.2. Actuators

The presented work aims at developing the faster SMA actuators. A series of experiments were conducted on a high-temperature (90°C transformation temperature) Flexinol® SMA actuator having 0.15 mm diameter. The length of SMA actuator while testing in ambient air was 80 mm; however, the length of SMA actuator in heat sink was 90 mm. The Flexinol® SMA actuators can withstand a maximum strain of 8%, whereas 4% strain is considered to be safe for cyclic operations.

5.1.3. Tubing

In this research, the SMA actuator was run through the Polytetrafluoroethylene (PTFE) tube along with the coolant. The care was taken that SMA actuator was completely jacketed around with the coolant. PTFE tube, due to its favourable mechanical and chemical properties, was preferred over other tubing. Various chemical characteristics that make PTFE material superior are its resistance to corrosive reagents, non-solubility, long-term weatherability, non-adhesiveness and non-flammability, whereas its mechanical properties include stability at high temperatures, flexibility at low temperature and low coefficient of friction [39].

5.1.4. Coolants

In order to achieve faster cooling, the PTFE tube was initially filled with high-temperature synthetic base grease (NLGI-3). Later, the tests were conducted with the PTFE tube filled with oil (Shell Helix Hx3 20W-50 Mobil Oil).

5.2. Methods

The SMA actuator was actuated using Joule heating method, with different level of actuation current in ambient air, grease and oil. In Joule heating, the SMA element is heated taking advantage of the resistance offered by it to the passage of electrical current. When the current passes through the SMA element, power losses are produced which in result heat the SMA actuator. This method of heating can be described by Eq. (2) (heating power).

$$P = I^2 R \tag{2}$$

Here,

P = Electrical power to heat the SMA actuator,

I = Current across actuator, measured in Ampere (A) and

R = Resistance offered by SMA actuator to the flow of current, measured in Ohm (Ω).

For experiments in ambient air, bare SMA actuator was used. Therefore, close look on the actuation current was necessary to avoid over heating of the actuator. The maximum current in ambient air tests was set as per given specification of the SMA actuator manufacturer. However, the SMA actuator was run through the PTFE tube while testing in grease and oil. PTFE material is flexible at low temperatures and is stable at high temperatures, also it offers less friction. However, during preparation for test, the sealing of the coolant, i.e. grease or oil inside the PTFE tube was a serious issue. This issue was later overcome by inserting a smaller diameter PTFE tube in the bottoms of the main PTFE tube, carrying the coolant and SMA actuator. Since grease and oil worked as heat sink, SMA actuator was tested at higher actuation currents with negligible chances of overheating. **Figure 4** shows the schematic diagram of the SMA actuator in PTFE tube along with the coolant.

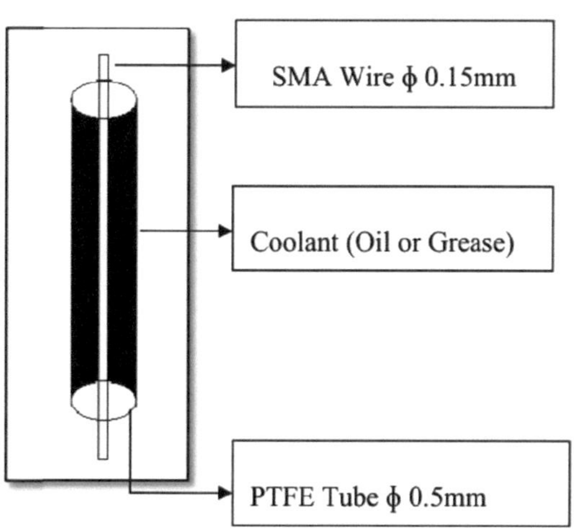

Figure 4. Proposed heat sink.

The SMA actuator was made to lift up a dead weight of 180 g, which induced 100 MPa stress in the SMA actuator. The data acquisition rate was set at 5 samples per second, which was sufficient to thoroughly examine the condition of SMA actuator during actuation and after the current supply was cutoff.

The thermal conductivity values of grease and oil are higher than the air (i.e. $\lambda_{OIL, GREASE} > \lambda_{AIR}$) as given in **Table 2**; therefore, the proposed heat sink will result in faster SMA cooling as compared to ambient air cooling.

Other factor affecting the cooling time is the surface area. Since the surface area of SMA is smaller than the surface area of PTFE tube (i.e. $A_{SMA} < A_{PTFE}$), the heat from the SMA will spread to the greater surface area of PTFE tube through the thermal conductive oil/grease, as shown in **Figure 5**.

Material	SMA	Teflon tubing	Thermal grease	Thermal oil	Air
Thermal conductivity (λ) in W m^{-1} K^{-1}	18 [3]	0.25 [40]	0.79 [41]	0.145 [40]	0.0257 [40]

Table 2. Thermal conductivity of different material.

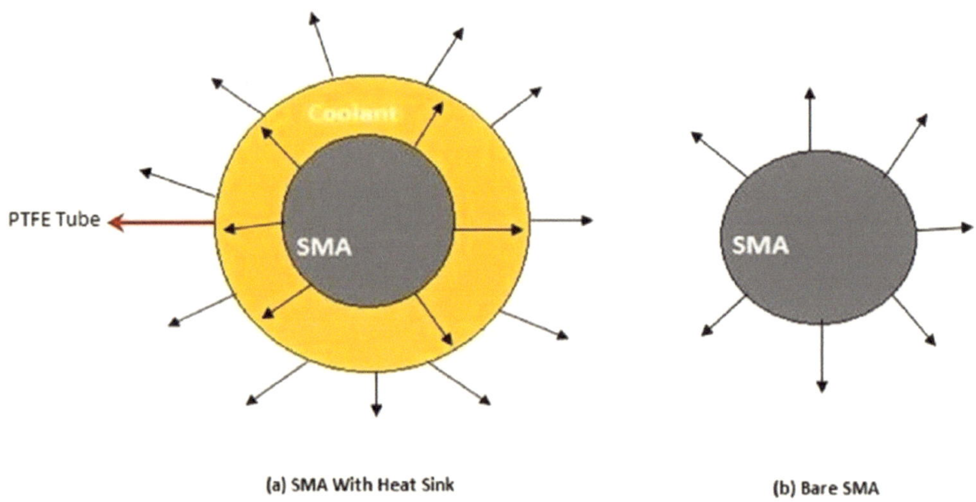

Figure 5. Area and heat dissipation model of (a) SMA with sink and (b) bare SMA.

6. Experiments and results

The SMA actuator was tested in ambient air, grease and oil, and time required by these cooling media was analysed. However, the time required to restore last 0.5% strain is not included, as it is not considered by the manufacturer. To assure the SMA properties, fresh specimen was used in each test. The strain induced in the SMA actuator is calculated by using Eq. (3) (strain induced in SMA actuator).

$$\%\ Strain = \frac{\Delta l}{l} \times 100 \tag{3}$$

Here,

Δl = Change in length of SMA actuator due to heating, taken in mm

l = Pre-deformed length of SMA actuator, taken in mm

6.1. Tests in ambient air

An 80 mm long bare SMA actuator specimen was tested in ambient air at 210, 310 and 410 mA actuation currents. The time for a cycle was set 10 seconds, of them 1 second for heating and remaining 9 seconds for relaxing of actuator. It was assured that SMA actuator should return to its original condition before the next cycle started. The response of SMA actuator in the ambient air actuation is given in **Figure 6**.

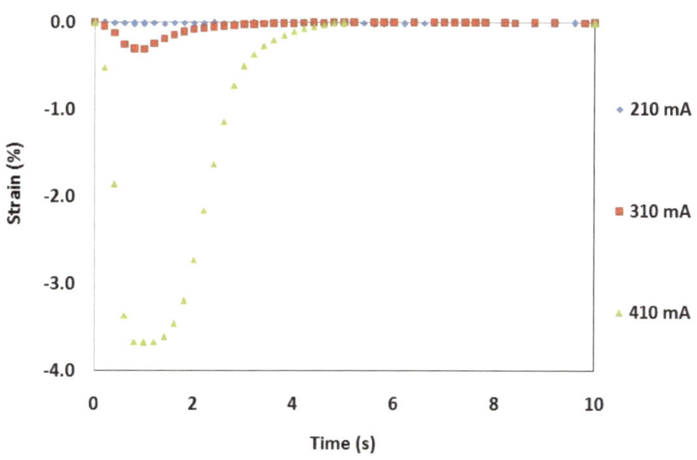

Figure 6. SMA actuator response in ambient air.

The negative sign on vertical axis in **Figure 6** shows the contraction in SMA actuator. It was found that heat energy produced at 210 mA heating current was too small to transform the SMA from martensite to austenite, hence resulting in negligible strain. When SMA actuator was actuated with 310 mA actuation current in ambient air, it could only be strained by 0.311%. This amount of strain is too small to be considered for any practical applications. The reason for small strain was the insufficient heat to cause a crystal rearrangement in the crystalline structure of the SMA. However, when SMA actuator was heated with 410 mA actuation current, a strain of 3.68%, i.e. 3 mm of contraction in SMA actuator was observed. This amount of strain is sufficient for many applications including minimal invasive surgery and other robotic applications. Upon the cutoff of power supply, the wire relaxed in 2 seconds. In such circumstances, the SMA actuator will have the working frequency of 0.333 Hz.

6.2. Tests in grease

For tests in grease, 90 mm long SMA actuator specimen was passed through an 80 mm long grease-filled PTFE tube. Being surrounded with grease, the chances of overheating were minimum; therefore, SMA actuator was actuated for 2 seconds as well. The SMA actuator sample was actuated with 210, 310, 410, 500, 615 and 720 mA for 1 and 2 seconds, whereas 810 mA actuation current was supplied for 1 second only.

6.2.1. One second actuation in grease

The SMA actuator specimen in grease-filled PTFE tube was actuated for 1 second at 210, 310, 410, 500, 615, 720 and 810 mA actuation currents. The time for a cycle was set 10 seconds, of them 1 second for heating and remaining 9 seconds for relaxing of actuator. It was assured that SMA could fully recover its initial form before the next cycle starts. **Figure 7** shows the response of SMA against 1 second actuation in grease.

Negative sign on the vertical axis in **Figure 7** shows contraction in the SMA actuator against the actuation.

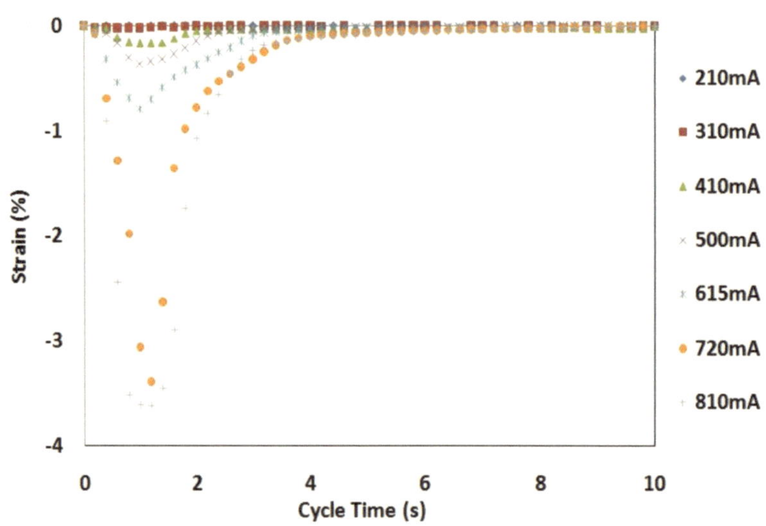

Figure 7. SMA actuator response against 1 second actuation in grease.

It can be observed that no strain was induced in the SMA actuator at 210 mA actuation current. The heat produced during actuation at 210 mA actuation was too low to cause a phase transformation in the actuator. At 310 mA heating current, the SMA actuator was strained by 0.0167%, which is too small for any application of SMA as actuator. At 410 mA heating current, due to rapid loss of heat to the coolant, i.e. grease, during heating, little amount of strain was induced in the SMA actuator, as compared to that produced in ambient air cooling, when heated with same level of actuation current. The actuation current of 410 mA could strain the grease-cooled actuator by 0.167%, which is only 4.54% of the strain produced in the ambient air-cooled SMA actuator at same actuation current. As discussed in the previous portions that due to higher thermal conductivity of grease than that of air, the heat loss rate to the grease was higher. Other reason for the rapid heat loss was the greater surface area of grease-cooled actuator, as it also included PTFE tube. The 500 and 615 mA actuation current resulted to 0.3635 and 0.7921% strain, respectively. Dynolloy Inc, in their technical data sheet for Flexinol SMA actuators, does not consider the time required to restore last 0.5% strain [3]. Therefore, in the grease-cooled SMA actuators, the cooling time for smaller strains is not considered. The SMA actuator deformed up to 3.047 mm when heated with 720 mA actuation current, this will induce a strain of 3.4%. This amount of strain is enough for SMAs to be applied as actuators in several applications like robotics, systems for drug delivery, catheters, etc. The SMA actuator cooled in 1.2 seconds. Sufficient quantity of heat was induced in the SMA to cause a phase transformation, when actuated with 810 mA actuation current. The heating was too fast for grease to absorb during actuation, due to which significant amount of heat was available inside the system to heat the SMA wire and cause a phase transformation. The 810 mA actuation current expanded the SMA actuator by 3.257 mm, resulting to 3.62% strain. However, the wire cooled off in 1.4 seconds, which is 30% improvement in cooling time as compared to that in ambient air. It should be noted that SMA actuator was strained by 3.68% when actuated with 410 mA actuation current in ambient air and recovered in 2 seconds.

6.2.2. *Two seconds actuation in grease*

The grease-cooled SMA actuators were actuated for 2 seconds with 210, 310, 410, 500, 615 and 720 mA actuation currents. These tests will help understand the effect of actuation time on the cooling efficiency of a heat sink. In these tests, the time for a cycle was set 10 seconds, of them 2 seconds for heating and remaining 8 seconds for relaxing of SMA actuator, in order to be assured that SMA fully recovers its pre-deformed shape and form before the next cycle starts. **Figure 8** shows the response of grease-cooled SMA actuator when actuated for 2 seconds at different actuation currents.

The negative sign on vertical axis represents contraction in SMA actuator, which is caused by heating.

It was observed that 210 mA actuation could not result in any deformation in SMA actuator, whereas a negligible amount of strain, about 0.04%, was observed in the SMA actuator at 310 mA actuation current. The heat energy produced by 210 and 310 mA actuation currents was too low to cause any phase transformation in the actuators.

When heated with 410 and 500 mA actuation currents, the SMA wire was strained by 0.35 and 0.6%, respectively. Because of rapid heat loss to the grease during heating, insufficient quantity of heat remained inside the SMA actuator to cause a complete phase transformation from martensite to the austenite. Whereas heating with 615 mA actuation current was sufficient to cause a significant amount of phase transformation in the SMA actuator. This could strain the actuator up to 3.2%, which is around four times higher than the strain produced by same level of actuation current when supplied for 1 second. The SMA wire actuator cooled off in 2.5 seconds, which is too high. The SMA actuator

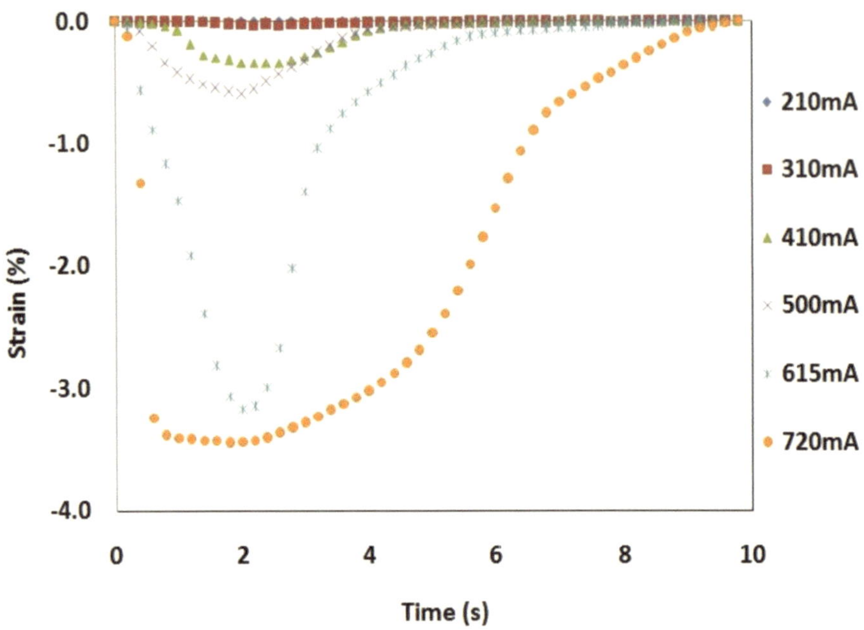

Figure 8. SMA actuator response against 2 seconds actuation in grease.

contracted up to 3.1 mm when heated with 720 mA actuation current supplied for 2 seconds. This resulted in 3.44% strain in SMA actuator, which is nearly equivalent to the strain produced with similar amount of actuation current when supplied for 1 second. However, the SMA wire cooled off in 6 seconds, which is five times greater than that required in 1 second actuation at same level of actuation current. This increase in cooling time shows that the efficiency of grease is affected with increasing heating time, especially at high actuation currents.

6.3. Tests in oil

To find out the effect of oil on SMA actuator cooling time, a 90 mm long SMA actuator specimen was run through an 80 mm long PTFE tube. The SMA actuator was heated with 210, 310, 410, 500, 615 and 720 mA current supplied for 1 and 2 seconds, whereas 810 mA actuation current was supplied for 1 second only.

6.3.1. One second actuation in oil

The SMA actuator specimen in oil-filled PTFE tube was actuated for 1 second at 210, 310, 410, 500, 615, 720 and 810 mA actuation currents. The time for a cycle was set 10 seconds, of them 1 second for heating and remaining 9 seconds for relaxing of SMA actuator. It was assured that SMA could fully recover its initial form before the next cycle starts. **Figure 9** shows the response of SMA against 1 second actuation in oil.

It was observed that 210 mA did not result in any strain in the SMA actuator; however, negligible strain in SMA actuator was observed at 310 and 410 mA actuation currents due to

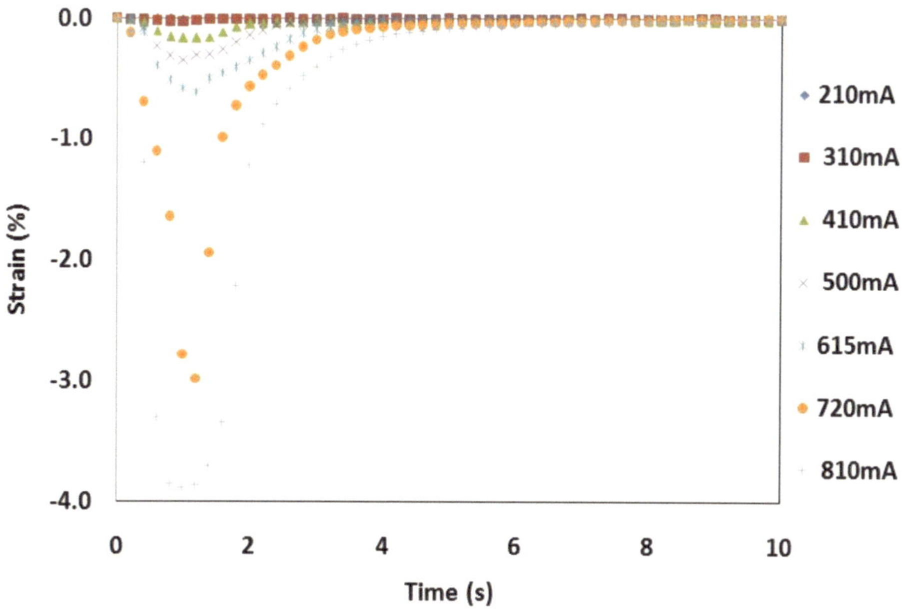

Figure 9. SMA actuator response against 1 second actuation in oil.

significant amount of heat loss to the oil during actuation. At 500 and 615 mA actuations, the SMA actuators strained by only 0.3472 and 0.6076%, respectively, due to insufficient amount of heat to cause phase transformation.

The 720 mA actuation current could produce sufficient heat to cause the phase transformation in the oil-covered SMA actuator. The 720 mA strained the SMA actuator up to 3%, producing 2.685 mm stroke length, which is sufficient for many SMA applications including catheters and latches and micro robots. The SMA actuator relaxed in 1 second hence capable to give 0.5 Hz frequency. While actuated with 810 mA current supply, the SMA was deformed by 3.5 mm, inducing 3.9%, which is 0.28% more than that produced with same actuation current in grease-cooled actuator. However, the cooling time required for SMA actuator was found to be 1.6 seconds.

6.3.2. Two seconds actuation in oil

The oil-cooled SMA actuators were actuated for 2 seconds with 210, 310, 410, 500, 615 and 720 mA actuation currents. SMA actuator heating for 2 seconds at various actuation currents will help understand the SMA behaviour against higher actuation times. These tests will also help understand the effect of actuation time on the cooling efficiency of a heat sink. In these set of tests, the time for a cycle was set 10 seconds, of them 2 seconds for heating and remaining 8 seconds for relaxing of SMA actuator. **Figure 10** shows the response of oil-cooled SMA actuator when actuated for 2 seconds at different actuation currents.

The negative sign on vertical axis in **Figure 10** denotes the shrinking in SMA actuator produced by heating.

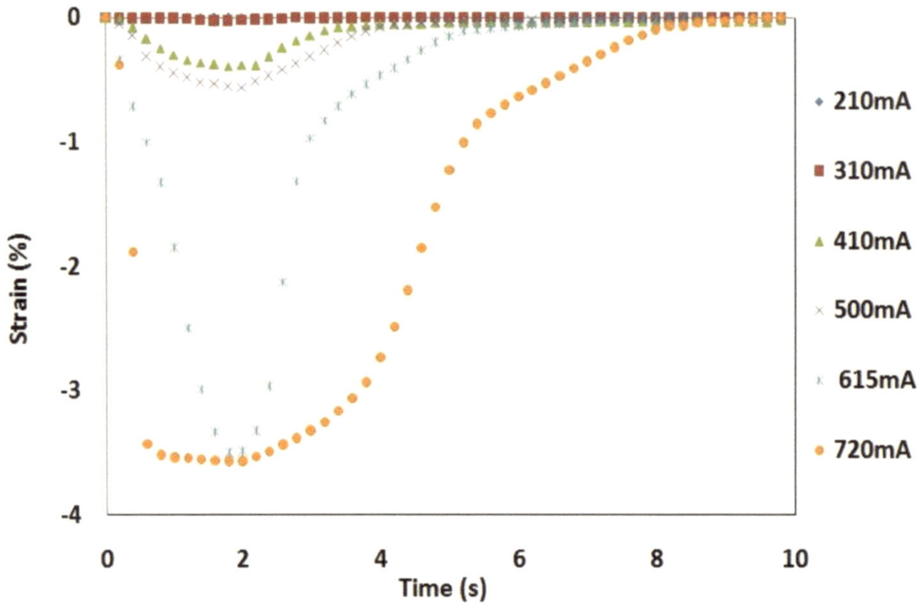

Figure 10. SMA actuator response against 2 seconds actuation in oil.

At 210 mA actuation current, no deformation in the SMA actuator was observed hence no strain in the SMA was induced, whereas negligible strain equal to 0.027% was observed at 310 mA actuation current. When actuated with 410 and 500 mA actuation currents, little strain equal to 0.4 and 0.56%, respectively, was observed. The observed quantity of strain is too small to for any real-time application of SMAs as actuators. However, 615 mA actuation current resulted in 3.5% strain in SMA, which is adequate for a number of real-world applications. The SMA specimen relaxed in 2 seconds, hence capable to give 0.33 Hz frequency. The SMA actuator contracted up to 3.213 mm, hence inducing 3.57% strain in the actuator, when heated with 720 mA actuation current. The actuator cooled off in 4.4 seconds. However, 410 mA actuation current in ambient air produced same amount of strain in the SMA actuator, whereas the wire had cooled off in 2 seconds, which is only 45.45% of cooling time required by SMA actuator when heated with 720 mA actuation current for 2 seconds in oil. The reason for the increase in cooling time is that the rejection of considerable amount of heat to the oil, which increased the temperature of oil, hence reducing the heat flow rate from actuator to the oil. It is general consideration that the heat transfer rate is proportional to difference of temperature between the two mediums.

7. Conclusions

From the results, it is derived that at same level of actuation current, grease- and oil-cooled SMA actuators underwent smaller strain as compared to the strain produced in ambient air cooling. The maximum heating current in ambient air was 410 mA, which strained the SMA actuator by 3.68%; however, the wire relaxed in 2 seconds, whereas an equivalent amount of strain in grease-cooled actuators was produced at 720 and 810 mA heating currents, and SMA actuators relaxed in 1.2 and 1.4 seconds, respectively, which is 40 and 30% improvement in cooling time of SMA as compared to that in ambient air. When actuated for 2 seconds in grease at 720 mA current supply, the actuator strained by 3.4% but recovered in 6 seconds, which is five times higher than the cooling time observed against 1 second actuation at 720 mA current in grease. Similarly, oil-cooled actuators produced considerable strain against 720 and 810 mA actuation current when heated for 1 second. The cooling times in this case were 1 and 1.6 seconds, respectively. Like grease-cooled actuators, the oil-cooled actuators required long cooling time against 2 seconds actuations, as the temperature of coolants, i.e. grease or oil was too high to efficiently absorb the heat. However, in case of ambient air cooling, close watch over the temperature of SMA is compulsory to prevent overheating which is likely to cause a permanent deformation in the actuator. However, in grease and oil cooling, the heat was rapidly being rejected to the coolant (i.e. grease or oil); therefore, likelihood of overheating of the SMA actuator is less. From the results and discussions sections, it can be observed that the oil-cooled actuator produced linear response, whereas the response of grease-cooled SMA actuator fluctuated at various points. The oil-cooled actuators produced more stroke length as compared to grease-cooled actuators and cooled off in less time. Therefore, for more strain, less cooling time, linear behaviour and safe actuation, the oil cooling is suggested over grease and ambient air cooling.

The presented research also reveals that increasing the actuation time also results in increased cooling time with no significant effect on the stroke length (i.e. strain). Therefore, it is further concluded that rather than increasing the heating duration with smaller magnitude currents, the magnitude of the heating current should be increased in order to avoid the heating of the sink.

Author details

Atta Muhammad Nizamani[1]*, Jawaid Daudpoto[1] and Muhammad Ali Nizamani[2]

*Address all correspondence to: attam.nizamani@gmail.com

1 Mehran University of Engineering and Technology, Jamshoro, Pakistan

2 Isra University, Hyderabad, Pakistan

References

[1] Ölander A. An electrochemical investigation of solid cadmium-gold alloys. Journal of the American Chemical Society. 1932;**54**:3819-3833

[2] Featherstone R, Teh YH. "Improving the Speed of Shape Memory Alloy Actuators by Faster Electrical Heating." Experimental Robotics IX, Springer (2006): 67-76. https://link.springer.com/chapter/10.1007%2F11552246_7?LI=true

[3] Dynalloy I. Technical Characteristics of Flexinol Actuator Wires. Dyanalloy Inc. Tustin, CA: 2011. www.dynalloy.com/pdfs/TCF1140.pdf

[4] Daudpoto, Jawaid, Ali Asghar Memon, and Imtiaz Hussain. "Actuation Characteristics of 0.15 Mm Diameter Flexinol® and Biometal® Wire Actuators for Robotic Applications." Mehran University of Research Journal of Engineering and Tecnology 32 (2013). Print.

[5] Lara-Quintanilla A, Bersee HE. A study on the contraction and cooling times of actively cooled shape memory alloy wires. Journal of Intelligent Material Systems and Structures 27.3 (2016): 403-17

[6] Nizamani, AM, J. Daudpoto, and MA Soomro. "Improving Cycle Time of Ni-Ti Shape Memory Alloy Actuators Using a Novel Cooling Technique." Sindh University Research Journal-SURJ (Science Series) 48.4D (2016), p 161-164

[7] Ikuta K. Micro/miniature shape memory alloy actuator. In: Proceedings, 1990 IEEE International Conference on Robotics and Automation, 1990; 13-18 May 1990; Cincinnati, Ohio, United States of America. IEEE; 1990. pp. 2156-2161

[8] Jani JM, Leary M, Subic A, Gibson MA. A review of shape memory alloy research, applications and opportunities. Materials & Design. 2014;**56**:1078-1113

[9] Butera F, Coda A, Vergani G, SpA SG. Shape memory actuators for automotive applications. Nanotec IT newsletter. Roma: AIRI/Nanotec IT; 2007. pp. 12-16

[10] Jani JM, Leary M, Subic A. Shape memory alloys in automotive applications. Applied Mechanics & Materials. 2014;**663**:248-253

[11] Zhong Z, Yeong C. Development of a gripper using SMA wire. Sensors and Actuators A: Physical. 2006;**126**:375-381

[12] Esfahani ET. Developing an Active Ankle foot Orthosis based on Shape Memory Alloys. The University of Toledo, Ohio, USA; 2007

[13] Copaci D, Flores A, Rueda F, Alguacil I, Blanco D, Moreno L. Wearable elbow exoskeleton actuated with shape memory alloy. In: Converging Clinical and Engineering Research on Neurorehabilitation II. Springer; 2017. pp. 477-481. Proceedings of the 3rd International Conference on NeuroRehabilitation (ICNR2016), October 18-21, 2016, Segovia, Spain. https://link.springer.com/chapter/10.1007/978-3-319-46669-9_79

[14] Rossi C, Colorado J, Coral W, Barrientos A. Bending continuous structures with SMAs: A novel robotic fish design. Bioinspiration & Biomimetics. 2011;**6**:045005

[15] Gaissert N, Mugrauer R, Mugrauer G, Jebens A, Jebens K, Knubben EM. Inventing a micro aerial vehicle inspired by the mechanics of dragonfly flight. 14th International Conference Towards Autonomous Robotic Systems,TAROS 2013, Oxford, UK, August 28--30, 2013. pp. 90-100. Springer, ISBN: 978-3-662-43645-5

[16] Coral Cuellar W, Rossi C, Colorado Montaño J, Barrientos Cruz A. SMA-Based MuscleLike Actuation in Biologically Inspired Robots: A State of the Art Review. Intech; 2012

[17] Taylor F, Au C. Forced air cooling of shape memory alloy actuators for a prosthetic hand. Journal of Computing and Information Science in Engineering (2016):16(4), 041004 (Nov 07, 2016) (5 pages) Paper No: JCISE-16-1051; doi:10.1115/1.4033233, Received January 29, 2016; Revised March 16, 2016 https://computingengineering.asmedigitalcollection.asme.org/article.aspx?articleID=2511361

[18] Morgan N. Medical shape memory alloy applications—the market and its products. Materials Science and Engineering: A. 2004;**378**:16-23

[19] Datla NV, Honarvar M, Nguyen TM, Konh B, Darvish K, Yu Y, et al. Towards a nitinol actuator for an active surgical needle. In: ASME 2012 Conference on Smart Materials, Adaptive Structures and Intelligent Systems; 19-21 September 2012; Georgia, USA. 2012. pp. 265-269. http://proceedings.asmedigitalcollection.asme.org/proceeding.aspx?articleid=1720812

[20] Guo S, Fukuda T, Kosuge K, Arai F, Oguro K, Negoro M. Micro catheter system with active guide wire-structure, experimental results and characteristic evaluation of active guide wire catheter using ICPF actuator. In: Proceedings, 1994 5th International Symposium on Micro Machine and Human Science; 2-4 October 1994; Cincinnati, OH, USA. IEEE; 1994. p. 191

[21] Reynaerts D, Peirs J, Van Brussel H. Design of a SMA-actuated implantable drug delivery system. In: MHS'95, Proceedings of the 6th International Symposium on Micro Machine and Human Science, 1995; 4-6 October 1995; Nagoya, Japan. IEEE; 1995. p. 111

[22] Potnuru A, Wu L, Tadesse Y. Artifiial heart for humanoid robot. Proc. SPIE 9056, Electroactive Polymer Actuators and Devices (EAPAD) 2014, 90562F (March 8, 2014); San Diego, California, USA doi:10.1117/12.2045289; http://dx.doi.org/10.1117/12.2045289

[23] Duerig T, Pelton A, Stöckel D. An overview of nitinol medical applications. Materials Science and Engineering: A. 1999;**273**:149-160

[24] Kudva JN. Overview of the DARPA smart wing project. Journal of Intelligent Material Systems and Structures. 2004;**15**:261-267

[25] Quan D, Hai X. Shape memory alloy in various aviation field. Procedia Engineering. 2015;**99**:1241-1246

[26] James H. Mabe ; Frederick T. Calkins and Mehmet B. Alkislar "Variable area jet nozzle using shape memory alloy actuators in an antagonistic design", Proc. SPIE 6930, Industrial and Commercial Applications of Smart Structures Technologies 2008, San Diego, California 69300T (March 19, 2008); doi:10.1117/12.776816; http://dx.doi.org/10.1117/12.776816

[27] Lucy MH, et al. "Report on Alternative Devices to Pyrotechnics on Spacecraft." (1996). 10th Annual AIAA/USU Conference on Small Satellites; 17-19 Sep. 1996; Logan, UT; United States. https://ntrs.nasa.gov/search.jsp?R=19960054342

[28] Allegranza, C., Gaillard, L., Le Letty, R., Patti, S., Scolamiero, L., & Toso, M. (2014, March). Actuators for Space Applications: State of the Art and New Technologies.

Actuator. In the proceedings of the 14th International Conference on New Actuators, Messe Bremen WFB Wirtschaftsförderung Bremen GmbH, Bremen, Germany. 742 p, ISBN 978-3-933339-22-5 . Proceedings available online at, http://www.actuator.de/userfiles/File/ACTUATOR_14/ACTUATOR_2014_ordform.pdf

[29] Olivier J. Godard ; Magdalini Z. Lagoudas and Dimitris C. Lagoudas "Design of space systems using shape memory alloys", Proc. SPIE 5056, Smart Structures and Materials 2003: Smart Structures and Integrated Systems, 545 (August 5, 2003) Volume 5056, San Diego, California, USA. http://dx.doi.org/10.1117/12.483469

[30] Hartl DJ, Lagoudas DC. Aerospace applications of shape memory alloys. Proceedings of the Institution of Mechanical Engineers, Part G: Journal of Aerospace Engineering. 2007; **221**:535-552

[31] Sonawane NS, Thangavelautham J. Precision pointing of antennas in space using arrays of shape memory alloy based linear actuators. arXiv preprint arXiv:1701.07561, 2017

[32] Loh CS, Yokoi H, Arai T. Natural heat-sinking control method for high-speed actuation of the SMA. International Journal of Advanced Robotic Systems. 2006;**3**:42

[33] Loh CS, Yokoi H, Arai T. New shape memory alloy actuator: Design and application in the prosthetic hand. In: IEEE-EMBS 2005. 27th Annual International Conference of the Engineering in Medicine and Biology Society, 2005; 17-18 January 2006; Shanghai, China. IEEE; 2006. pp. 6900-6903

[34] Cheng SS, Desai JP. Towards high frequency actuation of SMA spring for the neurosurgical robot-MINIR-II. In: 2015 IEEE International Conference on Robotics and Automation. (ICRA); 26-30 May 2015; Seattle, Washington, USA. IEEE; 2015. pp. 2580-2585

[35] Pathak A, Brei D, Luntz J. Transformation strain based method for characterization of convective heat transfer from shape memory alloy wires. Smart Materials and Structures. 2010;**19**:035005

[36] Tadesse Y, Thayer N, Priya S. Tailoring the response time of shape memory alloy wires through active cooling and pre-stress. Journal of Intelligent Material Systems and Structures. 2010;**21**:19-40

[37] Russell RA, Gorbet RB. Improving the response of SMA actuators. In: Proceedings, 1995 IEEE International Conference on Robotics and Automation, 1995; 21-27 May 1995; Nagoya, Japan. IEEE; 1995. pp. 2299-2304

[38] Soomro, MA, J. Daudpoto, and AM Nizamani. "Design, Development and Validation of an Experimental Test Bed for SMA Actuators." Sindh University Research Journal-SURJ (Science Series) 48.4D (2016). p 39-44

[39] Teflon P, Resin PF. Properties Handbook. Washington: DuPont Fluoroproducts; 1996

[40] Box ET. Thermal Conductivity of Materials and Gases [Internet]. Available from: http://www.engineeringtoolbox.com/thermal-conductivity-d_429.html [Accessed: 25 April 2016]

[41] Thermalloy. Thermal Greases [Internet]. Available from: htt://www.aavid.com/productgroup/interface/greases [Accessed: 25 April 2016]